LONDON MATHEMATICAL SOCIETY STUDENT TEXTS

Managing Editor: Professor E.B. Davies, Department of Mathematics,
King's College, Strand, London WC2R 2LS, England

1 Introduction to combinators and lambda-calculus, J.R. HINDLEY & J.P.SELDIN

2 Building models by games, WILFRID HODGES

3 Local fields, J.W.S. CASSELS

4 An introduction to twistor theory, S.A. HUGGETT & K.P. TOD

5 Introduction to general relativity, L. HUGHSTON & K.P. TOD

6 Lectures on stochastic analysis: diffusion theory, DANIEL W. STROOCK

T0296778

London Mathematical Society Students Texts. 6

Lectures on Stochastic Analysis: Diffusion Theory

DANIEL W. STROOCK
Massachusetts Institute of Technology

The right of the University of Cambridge to print and sell all manner of books was granted by Henry VIII in 1534. The University has printed and published continuously since 1584.

CAMBRIDGE UNIVERSITY PRESS
Cambridge
London New York New Rochelle
Melbourne Sydney

CAMBRIDGE UNIVERSITY PRESS
Cambridge, New York, Melbourne, Madrid, Cape Town, Singapore, São Paulo, Delhi

Cambridge University Press
The Edinburgh Building, Cambridge CB2 8RU, UK

Published in the United States of America by Cambridge University Press, New York

www.cambridge.org
Information on this title: www.cambridge.org/9780521333665

© Cambridge University Press 1987

This publication is in copyright. Subject to statutory exception
and to the provisions of relevant collective licensing agreements,
no reproduction of any part may take place without the written
permission of Cambridge University Press.

First published 1987
Re-issued in this digitally printed version 2008

A catalogue record for this publication is available from the British Library

Library of Congress Cataloguing in Publication data
Stroock, Daniel W.
Lectures on stochastic analysis
(London Mathematical Society Student Texts; 6)
Includes index.
1. Diffusion processes. I. Title. II. Series.
QA274.75.S85 1987 519.2'33 86-20782

ISBN 978-0-521-33366-5 hardback
ISBN 978-0-521-33645-1 paperback

Contents

Introduction

These notes grow out of lectures which I gave during the fall semester of 1985 at M.I.T. My purpose has been to provide a reasonably self-contained introduction to some stochastic analytic techniques which can be used in the study of certain analytic problems, and my method has been to concentrate on a particularly rich example rather than to attempt a general overview. The example which I have chosen is the study of second order partial differential operators of parabolic type. This example has the advantage that it leads very naturally to the analysis of measures on function space and the introduction of powerful probabilistic tools like martingales. At the same time, it highlights the basic virtue of probabilistic analysis: the direct role of intuition in the formulation and solution of problems.

The material which is covered has all been derived from my book [S.&V.] (Multidimensional Diffusion Processes, Grundlehren #233, Springer-Verlag, 1979) with S.R.S. Varadhan. However, the presentation here is quite different. In the first place, the emphasis there was on generality and detail; here it is on conceptual clarity. Secondly, at the time when we wrote [S.&V.], we were not aware of the ease with which the modern theory of martingales and stochastic integration can be presented. As a result, our development of that material was a kind of hybrid between the classical ideas of K. Itô and J.L. Doob and the modern theory based on the ideas of P.A. Meyer, H. Kunita, and S. Watanabe. In these notes the modern theory is presented; and the result is, I believe, not only more general but also more understandable.

In Chapter I, I give a quick review of a few of the important facts about probability measures on Polish spaces: the existence of regular conditional probability distributions and the theory of weak convergence. The chapter ends with the introduction of Wiener measure and a brief discussion of some of the basic elementary properties of Brownian motion.

Chapter II starts with an introduction to diffusion theory via the classical route of transition probability functions coming from the fundamental solution of parabolic equations. At the end of the first section, an attempt is made to bring out the analogy between diffusions and the theory of integral curves of a vector field. In this way I have tried to motivate the formulation (made precise in Chapter III) of diffusion theory in terms of martingales, and, at the same time, to indicate the central position which martingales play in stochastic analysis. The rest of Chapter II is devoted to the elements of martingale theory and the development of stochastic integration theory. (The presentation here profitted considerably from the incorporation of some ideas which I learned in the lectures given by K. Itô at the opening session of the I.M.A. in the fall of 1985.)

In Chapter III, I formulate the martingale problem and derive some of the basic facts about its solutions. The chapter ends with a proof that the martingale problem corresponding to a strictly elliptic operator with bounded continuous coefficients is well-posed. This proof turns on an elementary fact about singular integral operators, and a derivation of this fact is given in the appendix at the end of the chapter.

I. Stochastic Processes and Measures on Function Space:

1. Conditional Probabilities and Transition Probability Functions:

We begin by recalling the notion of conditional expectation. Namely, given a probability space $(E,\mathcal{F},P)$ and a sub-σ-algebra $\mathcal{F}'$, the conditional expectation value $E^P[X|\mathcal{F}']$ of a function $X \in L^2(P)$ is that $\mathcal{F}'$-measurable element of $L^2(P)$ such that

$$\int_A X(\xi)P(d\xi) = \int_A E^P[X|\mathcal{F}'](\xi)P(d\xi),\ A \in \mathcal{F}'. \tag{1.1}$$

Clearly $E^P[X|\mathcal{F}']$ exists: it is nothing more or less than the projection of X onto the subspace of $L^2(P)$ consisting of $\mathcal{F}'$-measurable P-square integrable functions. Moreover, $E^P[X|\mathcal{F}'] \geq 0$ (a.s.,P) if $X \geq 0$ (a.s.,P). Hence, if X is any non-negative $\mathcal{F}$-measurable function, then one can use the monotone convergence theorem to construct a non-negative $\mathcal{F}'$-measurable $E^P[X|\mathcal{F}']$ for which (1.1) holds; and clearly, up to a P-null set, there is only one such function. In this way, one sees that for any $\mathcal{F}$-measurable X which is either non-negative or in $L^1(P)$ there exists a P-almost-surely unique $\mathcal{F}'$-measurable $E^P[X|\mathcal{F}']$satisfying (1.1). Because $X \longmapsto E^P[X|\mathcal{F}']$ is linear and preserves non-negativity, one might hope that for each $\xi \in E$ there is a $P_\xi \in \mathbf{M}_1(E)$ (the space of probability measures on $(E,\mathcal{F})$) such that $E^P[X|\mathcal{F}'](\xi) = \int X(\eta)P_\xi(d\eta)$. Unfortunately, this hope is not fulfilled in general. However, it is fulfilled when one imposes certain topological conditions on $(E,\mathcal{F})$. Our first theorem addresses this question.

(1.2) <u>Theorem</u>: Suppose that Ω is a Polish space (i.e. Ω is a topological space which admits a complete separable metrization), and that $\mathcal{A}$ is a sub σ-algebra of $\mathcal{B}_\Omega$ (the Borel field over Ω). Given $P \in \mathbf{M}_1(\Omega)$, there is an $\mathcal{A}$-measurable map $\omega \longmapsto P_\omega \in \mathbf{M}_1(\Omega)$ such that $P(A \cap B) = \int_A P_\omega(B)\, P(d\omega)$ for all $A \in \mathcal{A}$ and $B \in \mathcal{B}_\Omega$. Moreover, $\omega \longmapsto P_\omega$ is uniquely determined up to a P-null set $A \in \mathcal{A}$. Finally, if $\mathcal{A}$ is countably generated, then $\omega \longmapsto P_\omega$ can be chosen so that $P_\omega(A) = \chi_A(\omega)$ for all $\omega \in \Omega$ and $A \in \mathcal{A}$.

<u>Proof</u>: Assume for the present that Ω is compact, the general case is left for later (c.f. exercise (2.2) below).

Choose $\{\varphi_n: n \geq 0\} \subseteq C(\Omega)$ to be a linearly independent set of functions whose span is dense in $C(\Omega)$, and assume that $\varphi_0 \equiv 1$. For each $n \geq 0$, let ψ_k be a bounded version of $E^P[\varphi_k | \mathcal{A}]$ and choose $\psi_0 \equiv 1$. Next, let A be the set of ω's such that there is an $n \geq 0$ and $a_0, \ldots, a_n \in \mathbb{Q}$ (the rationals) such that $\sum_{m=0}^{n} a_m \varphi_m \geq 0$ but $\sum_{m=0}^{n} a_m \psi_m(\omega) < 0$, and check that A is an $\mathcal{A}$-measurable P-null set. For $n \geq 0$ and $a_0, \ldots, a_n \in Q$, define $\Lambda_\omega(\sum_{m=0}^{n} a_m \varphi_m) = E^P[\sum_{m=0}^{n} a_m \varphi_m]$ if $\omega \in A$ and $\sum_{m=0}^{n} a_m \psi_m(\omega)$ otherwise. Check that, for each $\omega \in \Omega$, Λ_ω determines a unique non-negative linear functional on $C(\Omega)$ and that $\Lambda_\omega(1) = 1$. Further, check that $\omega \longmapsto \Lambda_\omega(\varphi)$ is $\mathcal{A}$-measurable for each $\varphi \in C(\Omega)$. Finally, let P_ω be the measure on Ω associated with Λ_ω by the Riesz representation theorem and check that

$\omega \longmapsto P_\omega$ satisfies the required conditions.

The uniqueness assertion is easy. Moreover, since $P_\bullet(A) = \chi_A(\cdot)$ (a.s.,P) for each $A \in \mathcal{A}$, it is clear that, when $\mathcal{A}$ is countably generated, $\omega \longmapsto P_\omega$ can be chosen so that this equality holds for all $\omega \in \Omega$. Q.E.D.

Referring to the set-up described in Theorem(1.2), the map $\omega \longmapsto P_\omega$ is called a <u>conditional probability distribution of P given $\mathcal{A}$</u> (abbreviated by <u>c.p.d. of P|$\mathcal{A}$</u>). If $\omega \longmapsto P_\omega$ has the additional property that $P_\omega(A) = \chi_A(\omega)$ for all $\omega \in \Omega$ and $A \in \mathcal{A}$, then $\omega \longmapsto P_\omega$ is called a <u>regular c.p.d. of P|$\mathcal{A}$</u> (abbreviated by <u>r.c.p.d. of P|$\mathcal{A}$</u>).

(1.3) <u>Remark</u>: The Polish space which will be the center of most of our attention in what follows is the space $\Omega = C([0,\infty);\mathbb{R}^N)$ of continuous paths from $[0,\infty)$ into $\mathbb{R}^N$ with the topology of uniform convergence on compact time intervals. Letting $x(t,\omega) = \omega(t)$ denote the position of $\omega \in \Omega$ at time $t \geq 0$, set $\mathcal{M}_t = \sigma(x(s): 0 \leq s \leq t)$ (the smallest σ-algebra over Ω with respect to which all the maps $\omega \longmapsto x(s,\omega)$, $0 \leq s \leq t$, are measurable). Given $P \in \mathbf{M}_1(\Omega)$, Theorem (1.2) says that for each $t \geq 0$ there is a P-essentially unique r.c.p.d. $\omega \longmapsto P_\omega^t$ of $P|\mathcal{M}_t$. Intuitively, the representation $P = \int P_\omega^t \, P(d\omega)$ can be thought of as a fibering of P according to how the path ω behaves during the initial time interval $[0,T]$. We will be mostly concerned with P's which are <u>Markov</u> in the sense that for each $t \geq 0$ and $B \in \mathcal{B}_\Omega$ which is measurable with respect to $\sigma(x(s): s \geq t)$, $\omega \longmapsto P_\omega^t(B)$ depends P-almost surely only on $x(t,\omega)$ and not on $x(s,\omega)$ for $s < t$.

2. The Weak Topology:

(2.1) Theorem: Let Ω be a Polish space and let ρ be a metric on Ω for which (Ω,ρ) is totally bounded. Suppose that Λ is a non-negative linear functional on $U(\Omega,\rho)$ (the space of ρ-uniformly continuous functions on Ω) satisfying $\Lambda(1) = 1$. Then there is a (unique) $P \in \mathbf{M}_1(\Omega)$ such that $\Lambda(\varphi) = E^P[\varphi]$ for all $\varphi \in U(\Omega,\rho)$ if and only if for all $\epsilon > 0$ there is a $K_\epsilon \subset\subset \Omega$ ("$\subset\subset$" is used to abbreviate "compact subset of") with the property that $\Lambda(\varphi) \geq 1 - \epsilon$ whenever $\varphi \in U(\Omega,\rho)$ satisfies $\varphi \geq \chi_{K_\epsilon}$.

Proof: Suppose that P exists. Choose $\{\omega_k\}$ to be a countable dense subset of Ω and for each $n \geq 1$ choose N_n so that $P(\bigcup_1^{N_n} B(\omega_k,1/n)) \geq 1 - \epsilon/2^n$, where the balls $B(\omega,r)$ are defined relative to a complete metric on Ω. Set $K_\epsilon = \bigcap_n \overline{\bigcup_1^{N_n} B(\omega_k,1/n)}$. Then $K_\epsilon \subset\subset \Omega$ and $P(K_\epsilon) \geq 1 - \epsilon$.

Next, suppose that $\Lambda(\varphi) \geq 1 - \epsilon$ whenever $\varphi \geq \chi_{K_\epsilon}$. Clearly we may assume that K_ϵ increases with decreasing ϵ. Let $\overline{\Omega}$ denote the completion of Ω with respect to ρ. Then $\varphi \in U(\Omega,\rho) \longmapsto \overline{\varphi}$, the unique extention of φ to $\overline{\Omega}$ in $C(\overline{\Omega})$, is an isometry from $U(\Omega,\rho)$ onto $C(\overline{\Omega})$. Hence, Λ induces a unique $\overline{\Lambda} \in C(\overline{\Omega})^*$ such that $\overline{\Lambda}(\overline{\varphi}) = \Lambda(\varphi)$, and so there is a $\overline{P} \in \mathbf{M}_1(\overline{\Omega})$ such that $\Lambda(\overline{\varphi}) = E^{\overline{P}}[\overline{\varphi}]$, $\varphi \in U(\Omega,\rho)$. Clearly, $\overline{P}(\Omega') = 1$ where $\Omega' = \bigcup_{\epsilon>0} K_\epsilon$, and so $P(\Gamma) = \overline{P}(\Gamma\cap\Omega')$ determines an element of $\mathbf{M}_1(\Omega)$ with the required property. The uniqueness of P is obvious.

Q.E.D.

(2.2) <u>Exercise</u>: Using the preceding, carry out the proof of Theorem (1.2) when Ω is not compact.

Given a Polish space Ω, the <u>weak topology</u> on $\mathbf{M}_1(\Omega)$ is the topology generated by sets of the form

$$\{\nu: \ |\nu(\varphi) - \mu(\varphi)| < \epsilon\},$$

for $\mu \in \mathbf{M}_1(\Omega)$, $\varphi \in C_b(\Omega)$, and $\epsilon > 0$. Thus, the weak topology on $\mathbf{M}_1(\Omega)$ is precisely the relative topology which $\mathbf{M}_1(\Omega)$ inherits from the weak* topology on $C_b(\Omega)^*$.

(2.3) <u>Exercise</u>: Let $\{\omega_k\}$ be a countable dense subset of Ω. Show that the set of convex combinations of the point masses δ_{ω_k} with non-negative rational coefficients is dense in $\mathbf{M}_1(\Omega)$. In particular, conclude that $\mathbf{M}_1(\Omega)$ is separable.

(2.4) <u>Lemma</u>: Given a net $\{\mu_\alpha\}$ in $\mathbf{M}_1(\Omega)$, the following are equivalent:

i) $\mu_\alpha \longrightarrow \mu$;

ii) ρ is a metric on Ω and $\mu_\alpha(\varphi) \longrightarrow \mu(\varphi)$ for every $\varphi \in U(\Omega,\rho)$;

iii) $\overline{\lim_\alpha}\ \mu_\alpha(F) \le \mu(F)$ for every closed F in Ω;

iv) $\underline{\lim_\alpha}\ \mu_\alpha(G) \ge \mu(G)$ for every open G in Ω;

v) $\lim_\alpha \mu(\Gamma) = \mu(\Gamma)$ for every $\Gamma \in \mathcal{B}_\Omega$ with $\mu(\partial\Gamma) = 0$.

<u>Proof</u>: Obviously i) implies ii) and iii) implies iv) implies v). To prove that ii) implies iii), set

$$\varphi_\epsilon(\omega) = \rho(\omega,(F^{(\epsilon)})^c)/[\rho(\omega,F) + \rho(\omega,(F^{(\epsilon)})^c)],$$

where $F^{(\epsilon)}$ is the set of ω's whose ρ-distance from F is less than ϵ. Then, $\varphi_\epsilon \in U(\Omega,\rho)$, $\chi_F \le \varphi_\epsilon \le \chi_{F^{(\epsilon)}}$, and so:

$$\overline{\lim_{\alpha}}\ \mu_\alpha(F) \leq \lim_{\alpha} \mu_\alpha(\varphi_\epsilon) = \mu(\varphi_\epsilon) \leq \mu(F^{(\epsilon)}).$$

After letting $\epsilon \longrightarrow 0$, one sees that iii) holds.

Finally, assume v) and let $\varphi \in C_b(\Omega)$ be given. Noting that $\mu(a < \varphi < b) = \mu(a \leq \varphi \leq b)$ for all but at most a countable number of a's and b's, choose for a given $\epsilon > 0$ a finite collection $a_0 < \ldots < a_N$ so that $a_0 < \varphi < a_N$, $a_n - a_{n-1} < \epsilon$, and $\mu(a_{n-1} < \varphi < a_n) = \mu(a_{n-1} \leq \varphi \leq a_n)$ for $1 \leq n \leq N$. Then:

$$|\mu_\alpha(\varphi) - \mu(\varphi)| \leq$$

$$2\epsilon + 2\|\varphi\|_{C_b(\Omega)} \sum_1^N |\mu_\alpha(a_{n-1} < \varphi \leq a_n) - \mu(a_{n-1} < \varphi \leq a_n)|,$$

and so, by v), $\overline{\lim_{\alpha}}\ |\mu_\alpha(\varphi) - \mu(\varphi)| \leq 2\epsilon$. Q.E.D.

(2.5) <u>Remark</u>: $\mathbf{M}_1(\Omega)$ admits a metric. Indeed, let ρ be a metric on Ω with the property that (Ω,ρ) is totally bounded. Then, since $U(\Omega,\rho)$ is isometric to $C(\overline{\Omega})$, there is a countable dense subset $\{\varphi_n\}$ of $U(\Omega,\rho)$. Define

$$R(\mu,\nu) = \sum_1^\infty |\mu(\varphi_n) - \nu(\varphi_n)|/2^n(1 + \|\varphi_n\|_{C_b(\Omega)}).$$

Clearly R is a metric for $\mathbf{M}_1(\Omega)$, and so (in view of (2.3)) we now see that $\mathbf{M}_1(\Omega)$ is a separable metric space. Actually, with a little more effort, one can show that $\mathbf{M}_1(\Omega)$ is itself a Polish space. The easiest way to see this is to show that $\mathbf{M}_1(\Omega)$ can be embedded in $\mathbf{M}_1(\overline{\Omega})$ as a G_δ. Since $\mathbf{M}_1(\overline{\Omega})$ is compact, and therefore Polish, it follows that $\mathbf{M}_1(\Omega)$ is also Polish. In any case, we now know that convergence and sequential convergence are equivalent in $\mathbf{M}_1(\Omega)$.

(2.6) Theorem(Prokhorov & Varadarajan): A set $\Gamma \subseteq \mathbf{M}_1(\Omega)$ is relatively compact if and only if for each $\epsilon > 0$ there is a $K_\epsilon \subset\subset \Omega$ such that $\mu(K_\epsilon) \geq 1 - \epsilon$ for every $\mu \in \Gamma$.

Proof: First suppose that $\Gamma \subset\subset \mathbf{M}_1(\Omega)$. Given $\epsilon > 0$ and $n \geq 1$, choose for each $\mu \in \Gamma$ a $K_n(\mu) \subset\subset \Omega$ so that $\mu(K_n(\mu)) > 1 - \epsilon/2^n$ and set $G_n(\mu) = \{\nu\colon \nu(K_n(\mu)^{(1/n)}) > 1 - \epsilon/2^n\}$, where distances are taken with respect to a complete metric on Ω. Next, choose $\mu_{n,1}, \ldots, \mu_{n,N_n} \in \Gamma$ so that $\Gamma \subseteq \bigcup_{k=1}^{N_n} G_n(\mu_{n,k})$, and set $K = \bigcap_{n=1}^{\infty} \bigcup_{k=1}^{N_n} \overline{K_n(\mu_{n,k})^{(1/n)}}$. Clearly $K \subset\subset \Omega$ and $\mu(K) \geq 1 - \epsilon$ for every $\mu \in \Gamma$.

To prove the opposite implication, think of $\mathbf{M}_1(\Omega)$ as a subset of the unit ball in $C_b(\Omega)^*$. Since the unit ball in $C_b(\Omega)^*$ is compact in the weak* topology, it suffices for us to check that every weak* limit Λ of μ's from Γ comes from an element of $\mathbf{M}_1(\Omega)$. But $\Lambda(\varphi) \geq 1 - \epsilon$ for all $\varphi \in C_b(\Omega)$ satisfying $\varphi \geq \chi_{K_\epsilon}$, and so Theorem(2.1) applies to Λ.

Q.E.D.

(2.7) Example: Let $\Omega = C([0,\infty);E)$, where (E,ρ) is a Polish space and we give Ω the topology of uniform convergence on finite intervals. Then, $\Gamma \subseteq \mathbf{M}_1(\Omega)$ is relatively compact if and only if for each $T > 0$ and $\epsilon > 0$ there exist $K \subset\subset E$ and $\delta\colon (0,\infty) \longrightarrow (0,\infty)$, satisfying $\lim_{\tau\downarrow 0} \delta(\tau) = 0$, such that:

$$\sup_{P\in\Gamma} P(\{\omega\colon x(t,\omega) \in K,\ t \in [0,T],\ \text{and}\ \rho(x(t,\omega),x(s,\omega)) \leq \delta(|t-s|),\ s,t \in [0,T]\}) \geq 1 - \epsilon.$$

In particular, if ρ-bounded subsets of E are relatively compact, then it suffices that:

$$\overline{\lim_{R\to\infty}} \sup_{P\in\Gamma} P(\{\omega: \rho(x,x(0,\omega)) \le R \text{ and } \rho(x(t,\omega),x(s,\omega)) \le \delta(|t-s|),\ s,t \in [0,T]\}) \ge 1 - \epsilon$$

for some reference point $x \in E$.

The following basic real-variable result was discovered by Garsia, Rademich, and Rumsey.

(2.8) Lemma(Garsia et al.): Let p and Ψ be strictly increasing continuous functions on $(0,\infty)$ satisfying $p(0) = \Psi(0) = 0$ and $\lim_{t\to\infty} \Psi(t) = \infty$. For given $T > 0$ and $\varphi \in C([0,T];\mathbb{R}^N)$, suppose that:

$$\int_0^T\int_0^T \Psi(|\varphi(t) - \varphi(s)|/p(|t - s|))dsdt \le B < \infty.$$

Then, for all $0 \le s \le t \le T$:

$$|\varphi(t) - \varphi(s)| \le 8\int_0^{t-s} \Psi^{-1}(4B/u^2)p(du).$$

Proof: Define

$$I(t) = \int_0^T \Psi(|\varphi(t) - \varphi(s)|/p(|t - s|))ds,\ t \in [0,T].$$

Since $\int_0^T I(t)dt \le B$, there is a $t_0 \in (0,T)$ such that $I(t_0) \le B/T$. Next, choose $t_0 > d_0 > t_1 > \dots > t_n > d_n > \dots$ as follows. Given t_{n-1}, define d_{n-1} by $p(d_{n-1}) = 1/2p(t_{n-1})$ and choose $t_n \in (0,d_{n-1})$ so that $I(t_n) \le 2B/d_{n-1}$ and

$$\Psi(|\varphi(t_n) - \varphi(t_{n-1})|/p(|t_n-t_{n-1}|)) \le 2I(t_{n-1})/d_{n-1}.$$

Such a t_n exists because each of the specified conditions can fail on a set of at most measure $d_{n-1}/2$.

Clearly:

$$2p(d_{n+1}) = p(t_{n+1}) \le p(d_n).$$

Thus, $t_n \downarrow 0$ as $n \longrightarrow \infty$. Also, $p(t_n - t_{n+1}) \le p(t_n) = 2p(d_n) = 4(p(d_n) - 1/2p(d_n)) \le 4(p(d_n) - p(d_{n+1}))$. Hence, with $d_{-1} \equiv T$:

$$\begin{aligned} |\varphi(t_{n+1}) - \varphi(t_n)| &\le \Psi^{-1}(2I(t_n)/d_n)p(t_n - t_{n+1}) \\ &\le \Psi^{-1}(4B/d_{n-1}d_n)(p(d_n) - p(d_{n+1})) \\ &\le 4\int_{d_{n+1}}^{d_n} \Psi^{-1}(4B/u^2)p(du), \end{aligned}$$

and so $|\varphi(t_0) - \varphi(0)| \le 4\int_0^T \Psi^{-1}(4B/u^2)p(du)$. By the same argument going in the opposite time direction, $|\varphi(T) - \varphi(t_0)| \le 4\int_0^T \Psi^{-1}(4B/u^2)p(du)$. Hence, we now have:

$$|\varphi(T) - \varphi(0)| \le 8\int_0^T \Psi^{-1}(4B/u^2)p(du). \tag{2.9}$$

To complete the proof, let $0 \le \sigma \le \tau \le T$ be given and apply (2.9) to $\bar{\varphi}(t) = \varphi(\sigma + (\tau-\sigma)t/T)$ and $\bar{p}(t) = p((\tau-\sigma)t/T)$. Since

$$\begin{aligned} &\int_0^T\int_0^T \Psi(|\bar{\varphi}(t) - \bar{\varphi}(s)|/\bar{p}(|t - s|))dsdt = \\ &(T/(\tau-\sigma))^2\int_\sigma^\tau\int_\sigma^\tau \Psi(|\varphi(t) - \varphi(s)|/p(|t - s|))dsdt \\ &\le (T/(\tau-\sigma)^2B \equiv \bar{B}, \end{aligned}$$

we conclude that:

$$\begin{aligned} |\varphi(\tau) - \varphi(\sigma)| &\le 8\int_0^T \Psi^{-1}(4\bar{B}/u^2)\bar{p}(du) \\ &= 8\int_0^{\tau-\sigma} \Psi^{-1}(4B/u^2)p(du). \end{aligned}$$

Q.E.D.

(2.10)<u>Exercise</u>: Generalize the preceding as follows.

Let $(L,\|\cdot\|)$ be a normed linear space, $r > 0$, $d \in Z^+$, and $\varphi:\mathbb{R}^d \longrightarrow L$ a weakly continous map. Set $B(r) = \{x\in\mathbb{R}^d: |x| < r\}$ and suppose that

$$\int_{B(r)}\int_{B(r)} \Psi(\|\varphi(y) - \varphi(x)\|/p(|y-x|))dxdy \leq B < \infty.$$

Show that

$$\|\varphi(y) - \varphi(x)\| \leq 8\int_0^{|y-x|} \Psi^{-1}(4^{d+1}B/\gamma u^{2d})p(du)$$

where

$$\gamma = \gamma_d \equiv \inf\{|(x + B(r))\cap B(1)|/r^d: |x| \leq 1 \text{ and } r \leq 2\}.$$

A proof can be made by mimicking the argument used to prove Lemma(2.9) (cf. 2.4.1 on p. 60 of [S.&V.]).

(2.12)<u>Theorem</u> (Kolmogorov's Criterion): Let $(\Omega,\mathcal{F},P)$ be a probability space and ξ a measurable map of $\mathbb{R}^d\times\Omega$ into the normed linear space $(L,\|\cdot\|)$. Assume that $x \in R^d \longmapsto \xi(x,\omega)$ is weakly continuous for each $\omega \in \Omega$ and that for some $q \in [1,\infty)$, $r > 0$, $\alpha > 0$, and $A < \infty$:

$$E^P[\|\xi(y) - \xi(x)\|^q] \leq A|y - x|^{d+\alpha}, \quad x,y \in B(r). \tag{2.13}$$

Then, for all $\lambda > 0$,

$$P(\sup_{x,y\in B(r)}\|\xi(y) - \xi(x)\|/|y - x|^\beta \geq \lambda) \leq AB/\lambda^q \tag{2.14}$$

where $\beta = \alpha/2q$ and $B < \infty$ depends only on d, q, r, and α.

<u>Proof</u>: Let $\rho = 2d + \alpha/2$. Then:

$$\int_{B(r)}\int_{B(r)} E^P[(\|\xi(y) - \xi(x)\|/|y - x|^{\rho/q})^q)]dxdy \leq AB'$$

where

$$B' \equiv \int_{B(r)}\int_{B(r)} |y - x|^{-d+\alpha/2}dxdy.$$

Next, set

$$Y(\omega) = \int_{B(r)}\int_{B(r)} [\|\xi(y,\omega) - \xi(x,\omega)\|/|y - x|^{\rho/q}]^q dxdy.$$

Then, by Fubini's theorem, $E^P[Y] \le AB'$, and so:

$$P(Y \ge \lambda^q) \le AB'/\lambda^q, \ \lambda > 0.$$

In addition, by (2.10):

$$\|\xi(y,\omega) - \xi(x,\omega)\| \le 8\int_0^{|y-x|} (4^{d+1}Y(\omega)/\gamma u^{2d})^{1/q} du^{\rho/q}$$
$$\le CY(\omega)^{1/q}|y - x|^{\beta}.$$

Q.E.D.

(2.15) <u>Corollary</u>: Let $\Omega = C([0,\infty);\mathbb{R}^N)$ and suppose that $\Gamma \subseteq \mathbf{M}_1(\Omega)$ has the properties that:

$$\lim_{R\to\infty} \sup_{P\in\Gamma} P(|x(0)| \ge R) = 0$$

and that for each $T > 0$:

$$\sup_{P\in\Gamma} \sup_{0\le s<t\le T} E^P[|x(t) - x(s)|^q]/(t - s)^{1+\alpha} < \infty,$$

for some $q \in [1,\infty)$ and $\alpha > 0$. Then, Γ is relatively compact.

3. Constructing Measures on $C([0,\infty);\mathbb{R}^N)$:

Throughout this section, and very often in what follows, Ω denotes the Polish space $C([0,\infty);\mathbb{R}^N)$, $\mathcal{M} = \mathcal{B}_\Omega$, and $\mathcal{M}_t = \sigma(x(s): 0 \le s \le t)$, $t \ge 0$.

(3.1) Exercise: Check that $\mathcal{M} = \sigma(\bigcup_{t\ge 0} \mathcal{M}_t)$. In particular, conclude that if $P,Q \in \mathbf{M}_1(\Omega)$ satisfy $P(x(t_0) \in \Gamma_0,\ldots,x(t_n) \in \Gamma_n) = Q(x(t_0) \in \Gamma_0,\ldots,x(t_n) \in \Gamma_n)$ for all $n \ge 0$, $0 \le t_0 <\ldots < t_n$, and $\Gamma_0,\ldots,\Gamma_n \in \mathcal{B}_{\mathbb{R}^N}$, then $P = Q$.

Next, for each $n \ge 0$ and $0 \le t_0 \le\ldots\le t_n$, suppose that $P_{t_0,\ldots,t_n} \in \mathbf{M}_1((\mathbb{R}^N)^n)$ and assume that the family $\{P_{t_0,\ldots,t_n}\}$ is consistent in the sense that:

$$P_{t_0,\ldots,t_{k-1},t_{k+1},\ldots,t_n}(\Gamma_0\times\ldots\times\Gamma_{k-1}\times\Gamma_{k+1}\times\ldots\times\Gamma_n)$$
$$= P_{t_0,\ldots,t_n}(\Gamma_0\times\ldots\times\Gamma_{k-1}\times E\times\Gamma_{k+1}\times\ldots\times\Gamma_n) \tag{3.2}$$

for all $n \ge 0$, $0 \le k \le n$, $t_0<\ldots< t_n$, and $\Gamma_0,\ldots,\Gamma_n \in \mathcal{B}_{\mathbb{R}^N}$.

(3.3) Example: One of the most important sources of consistent families are (Markov) transition probability functions. Namely, the function $P(s,x;t,\cdot)$, defined for $0 \le s < t$ and $x \in \mathbb{R}^N$ and taking values in $\mathbf{M}_1(\mathbb{R}^N)$, is called a transition probability function on $\mathbb{R}^N$ if it is measurable and satisfies the Chapman-Kolmogorov equation:

$$P(s,x;u,\cdot) = \int_{\mathbb{R}^N} P(t,y;u,\cdot)P(s,x;t,dy) \tag{3.4}$$

for all $0 \le s < t$ and $x \in \mathbb{R}^N$. Given an initial distribution $\mu_0 \in \mathbf{M}_1(\mathbb{R}^N)$, we associate with μ_0 and $P(s,x;t,\cdot)$ the consistent family $\{P_{t_0,\ldots,t_n}\}$ determined by

$$P_{t_0,\ldots,t_n}(\Gamma_0\times\ldots\times\Gamma_n)$$
$$= \int_{\Gamma_0}\mu_0(dx_0)\int_{\Gamma_1}P(t_0,x_0;t_1,dx_1)\ldots\int_{\Gamma_n}P(t_{n-1},x_{n-1};t_n,dx_n).$$

(3.5) <u>Theorem</u>: Let $\{P_{t_0,\ldots,t_n}\}$ be a consistent family and assume that for each $T > 0$:

$$\sup_{0\leq s\leq t\leq T}\int |y - x|^q\, P_{s,t}(dx\times dy)/(t-s)^{1+\alpha} < \infty \tag{3.6}$$

for some $q \in [1,\infty)$ and $\alpha > 0$. Then there exists a unique $P \in \mathbf{M}_1(\Omega)$ such that $P_{t_0,\ldots,t_n} = P\circ(x(t_0),\ldots,x(t_n))^{-1}$ for all $n \geq 0$ and $0 \leq t_0 <\cdots< t_n$. (Throughout, $P\circ\Phi^{-1}(\Gamma) \equiv P(\Phi^{-1}(\Gamma))$.)

<u>Proof</u>: The uniqueness is immediate from exercise (3.1). To prove existence, define for $m \geq 0$ the map $\Phi_m:(\mathbb{R}^N)^{4^m+1}\longrightarrow\Omega$ so that $x(t,\Phi_m(x_0,\ldots,x_{4^m})) = x_k + 2^m(t - k/2^m)(x_{k+1} - x_k)$ if $k/2^m \leq t < (k+1)/2^m$ and $0 \leq k < 4^m$, and $x(t,\Phi_m(x_0,\ldots,x_{4^m})) = x_{4^m}$ if $t \geq 2^m$. Next, set $P_m = P_{t_0,\ldots,t_{n_m}}\circ\Phi_m^{-1}$ where $n_m = 4^m$ and $t_k = k/2^m$. Then, by Corollary (2.15), $\{P_m: m \geq 0\}$ is relatively compact in $\mathbf{M}_1(\Omega)$. Moreover, if P is any limit of $\{P_m: m \geq 0\}$, then

$$E^P[\varphi_0(x(t_o))\ldots\varphi_n(x(t_n))]$$
$$= \int \varphi_0(x_o)\ldots\varphi_n(x_n)P_{t_0,\ldots,t_n}(dx_0\times\cdots\times dx_n) \tag{3.7}$$

for all $n \geq 0$, dyadic $0 \leq t_0 <\ldots< t_n$, and $\varphi_0,\ldots,\varphi_n \in C_b^1(\mathbb{R}^N)$. Since both sides of (3.7) are continuous with respect to $(t_0,\ldots,t_n)$, it follows that P has the required property.

Q.E.D.

(3.8) <u>Exercise</u>: Use (3.6) to check the claims, made in the preceding proof, that $\{P_m : m \geq 0\}$ is relatively compact and that the right hand side of (3.7) is continuous with respect to $(t_0, \ldots, t_n)$. Also, show that if $P(s,x;t,\cdot)$ is a transition probability function which satisfies:

$$\sup_{0 \leq s < t \leq T} \int |y - x|^q \, P(s,x;t,dy)/(t-s)^{1+\alpha} < \infty \tag{3.9}$$

for each $T > 0$ and some $q \in [1,\infty)$ and $\alpha > 0$, then the family associated with any initial distribution and $P(s,x;t,\cdot)$ satisfies (3.6).

4. Wiener Measure, Some Elementary Properties:

We continue with the notation used in the preceding section. The classic example of a measure on Ω is the one constructed by N. Wiener. Namely, set $P(s,x;t,dy) = g(t-s,y-x)dy$, where

$$g(s,x) \equiv (2\pi)^{-N/2}\exp(-|x|^2/2s) \tag{4.1}$$

is the (standard) Gauss (or Weierstrass) kernel. It is an easy computation to check that:

$$\int \exp[\sum_{j=1}^{N} \theta_j y_j] g(t,y)dy = \exp[t/2 \sum_{j=1}^{N} \theta_j^2] \tag{4.2}$$

for any $t > 0$, and $\theta_1,\ldots,\theta_N \in \mathbb{C}$; and from (4.2), one can easily show that $P(s,x;t,\cdot)$ is a transition probability function which satisfies

$$\int |y - x|^q \, P(s,x;t,dy) = C_N(q)(t - s)^{q/2} \tag{4.3}$$

for each $q \in [1,\infty)$ In particular, (3.9) holds with $q = 4$ and $\alpha = 1$. The measure $P \in \mathbf{M}_1(\Omega)$ corresponding to an initial distribution μ_0 and this $P(s,x;t,.)$ is called the (N-dimensional) Wiener measure with initial distribution μ_0 and is denoted by $\mathcal{W}_{\mu_0}$. In particular, when $\mu_0 = \delta_x$, we use $\mathcal{W}_x$ in place of $\mathcal{W}_{\delta_x}$ and refer to $\mathcal{W}_x$ as the (N-dimensional) Wiener measure starting at x; and when $x = 0$, we will use $\mathcal{W}$ (or, when dimension is emphasized, $\mathcal{W}^{(N)}$) instead of $\mathcal{W}_0$ and will call $\mathcal{W}$ the (N-dimensional) Wiener measure. In this connection, we introduce here the notion of an N-dimensional Wiener process. Namely, given a probability space $(E,\mathcal{F},P)$, we will say that $\beta:[0,\infty)\times E \longrightarrow R^N$ is an (N-dimensional)-Wiener process under P if β is measurable, $t \longmapsto \beta(t)$ is P-almost surely continuous, and $P\circ\beta(\cdot)^{-1} = \mathcal{W}^{(N)}$.

(4.4) <u>Exercise</u>: Identifying $C([0,\infty);\mathbb{R}^N)$ with $C([0,\infty);\mathbb{R}^1)^N$, show that $\mathcal{W}^{(N)} = (\mathcal{W}^{(1)})^N$. In addition, show that $-x(\cdot)$ is a Wiener process under $\mathcal{W}$ and that $\mathcal{W}_x = \mathcal{W}\circ T_x^{-1}$ where $T_x\colon \Omega\longrightarrow\Omega$ is given by $x(t,T_x(\omega)) = x + x(t,\omega)$. Finally, for a given $s \geq 0$, let $\omega\longmapsto\mathcal{W}_\omega^s$ be the r.c.p.d. of $\mathcal{W}|\mathcal{M}_s$. Show that, for $\mathcal{W}$-almost all ω, $x(\cdot+s) - x(s,\omega)$ is a Wiener process under $\mathcal{W}_\omega^s$, and use this to conclude that $\mathcal{W}_\omega^s\circ\theta_s^{-1} = \mathcal{W}_{x(s,\omega)}$ (a.s.,$\mathcal{W}$), where <u>θ_s</u>: $\Omega\longrightarrow\Omega$ is the <u>time shift map</u> given by $x(\cdot,\theta_s\omega) = x(\cdot+s,\omega)$.

Thus far we have discussed Wiener measure from the Markovian point of view (i.e. in terms of transition probability functions). An equally important way to approach this subject is from the Gaussian standpoint. From the Gaussian standpoint, $\mathcal{W}$ is characterized by the equation:

$$
\begin{aligned}
E^{\mathcal{W}}\Big[\exp\Big(i\sum_{k=1}^{N}(\theta_k,x(t_k))_{\mathbb{R}^N}\Big)\Big] &= \\
&\exp\Big(-1/2\sum_{k,\ell=1}^{N}(t_k\wedge t_\ell)(\theta_k,\theta_\ell)_{\mathbb{R}^N}\Big),
\end{aligned}
\tag{4.5}
$$

for all $n \geq 1$, $t_1,\ldots,t_n \in [0,\infty)$, and $\theta_1,\ldots,\theta_n \in \mathbb{R}^N$.

(4.6) <u>Exercise</u>: Check that (4.5) holds and that it characterizes $\mathcal{W}$. Next, define $\Phi_\lambda\colon \Omega\longrightarrow\Omega$ by $x(t,\Phi_\lambda(\omega)) = \lambda^{1/2}x(t/\lambda,\omega)$ for $\lambda > 0$; and, using (4.5), show that $\mathcal{W} = \mathcal{W}\circ\Phi_\lambda^{-1}$. This invariance property of $\mathcal{W}$ is often called the <u>Brownian scaling</u> propery. In order to describe the <u>time inversion</u> property of Wiener processes one must first check that $\mathcal{W}(\lim_{t\longrightarrow\infty} x(t)/t = 0) = 1$. To this end, note that:

$$\mathcal{W}\left(\sup_{t\geq m}|x(t)|/t \geq \epsilon\right) \leq \sum_{n=m}^{\infty} \mathcal{W}\left(\sup_{n\leq t\leq n+1}|x(t)| \geq n\epsilon\right)$$

and that, by Brownian scaling:

$$\mathcal{W}\left(\sup_{n\leq t\leq n+1}|x(t)| \geq n\epsilon\right) \leq \mathcal{W}\left(\sup_{0\leq t\leq 2}|x(t)| \geq n^{1/2}\epsilon\right).$$

Now combine (4.3) with (2.14) to conclude that

$$\mathcal{W}\left(\sup_{n\leq t\leq n+1}|x(t)| \geq n\epsilon\right) \leq C/n^2\epsilon^4,$$

and therefore that $\mathcal{W}(\lim_{t\to\infty} x(t)/t = 0) = 1$. The Brownian time inversion property can now be stated as follows. Define $\beta(0) \equiv 0$ and, for $t > 0$, set $\beta(t) = tx(1/t)$. Using the preceding and (4.5), check that $\beta(\cdot)$ is a Wiener process under $\mathcal{W}$.

We close this chapter with a justly famous result due to Wiener. In the next chapter we will derive this same result from a much more sophisticated viewpoint.

(4.6) <u>Theorem</u>: $\mathcal{W}$-almost no $\omega \in \Omega$ is Lipschitz continuous at even one $t \geq 0$.

<u>Proof</u>: In view of exercise (4.4), it suffices to treat the case when $N = 1$ and to show that $\mathcal{W}$-almost no ω is Lipschitz continuous at any $t \in [0,1)$. But if ω were Lipschitz continuous at $t \in [0,1)$, then there would exist $\ell, m \in Z^+$ such that $|x((k+1/n)) - x(k/n)|$ would be less than ℓ/n for all $n \geq m$ and three consecutive k's between 0 and $(n+2)$. Hence, it suffices to show that the sets

$$B(\ell,m) = \bigcap_{n=m}^{\infty} \bigcup_{k=1}^{n} \bigcap_{j=0}^{2} A(\ell,n,k+j), \quad \ell,m \in Z^+,$$

where $A(\ell,n,k) \equiv \{|x((k+1/n)) - x(k/n)| \leq \ell/n\}$, have $\mathcal{W}$-measure 0. But, by (4.1) and Brownian scaling:

$$\mathcal{W}(B(\ell,m)) \le \overline{\lim_{n\to\infty}}\, n\mathcal{W}(|x(1/n)| \le \ell/n)^3$$
$$= \overline{\lim_{n\to\infty}}\, n\mathcal{W}(|x(1)| \le \ell/n^{1/2})^3$$
$$= \overline{\lim_{n\to\infty}}\, n\left(\int_{-\ell/n^{1/2}}^{\ell/n^{1/2}} g(1,y)dy\right)^3 = 0.$$

Q.E.D.

(4.7) Remark: P. Levy obtained a far sharper version of the preceding; namely, he showed that:

$$\mathcal{W}\left(\overline{\lim_{\delta\downarrow 0}} \sup_{\substack{0\le s<t\le 1\\ t-s<\delta}} |x(t) - x(s)|/(2\delta\log 1/\delta)^{1/2} = 1\right) = 1. \qquad (4.8)$$

The lower bound on the lim sup is a quite elementary application of the Borel-Cantelli lemma (cf. p. 14 of H. P. McKean's Stochastic Integrals, Academic Press, 1969), but the upper bound is a little difficult. A derivation of the less sharp estimate when the upper bound 1 is replaced by 8 can be based on the reasoning used to prove (2.14). See 2.4.8 in [S.&V.] for more details.

II. DIFFUSIONS AND MARTINGALES:

1. A Brief Introduction to Classical Diffusion Theory:

We continue with the notation used in section I.3.

Let $\underline{S}^{+}(\mathbb{R}^N)$ denote the space of non-negative definite symmetric matrices, and for given bounded measurable functions $a: [0,\infty)\times\mathbb{R}^N \longrightarrow S^{+}(\mathbb{R}^N)$ and $b: [0,\infty)\times\mathbb{R}^N \longrightarrow \mathbb{R}^N$ define the operator valued map $t \in [0,\infty) \longmapsto L_t$ by

$$L_t = 1/2 \sum_{i,j=1}^{N} a^{ij}(t,x)\partial_{x^i}\partial_{x^j} + \sum_{i=1}^{N} b^i(t,x)\partial_{x^i}. \tag{1.1}$$

The following theorem can be proved using quite elementary analytic methods (cf. Chapter 3 in [S.&V.]).

(1.2) Theorem: Assume that $a \in C_b^{0,3}([0,\infty)\times\mathbb{R}^N;S^{+}(\mathbb{R}^N))$ and that $b \in C_b^{0,2}([0,\infty)\times\mathbb{R}^N;\mathbb{R}^N)$. Then there is a unique transition probability function $P(s,x;t,\cdot)$ on $\mathbb{R}^N$ such that for each $T > 0$ and all $f \in C_b^{1,2}([0,T]\times\mathbb{R}^N)$:

$$\begin{aligned}&\int f(T,y)P(s,x;T,dy) - f(s,x)\\ &= \int_s^T dt\int(\partial_t + L_t)f(t,y)P(s,x;t,dy),\quad (s,x) \in [0,T]\times\mathbb{R}^N.\end{aligned} \tag{1.3}$$

Moreover, if $T > 0$ and $\varphi \in C_0^{\infty}(\mathbb{R}^N)$, then $(s,x) \in [0,T]\times\mathbb{R}^N \longmapsto u_{T,\varphi}(s,x) \equiv \int\varphi(y)P(s,x;T,dy)$ is an element of $C_b^{1,2}([0,T]\times\mathbb{R}^N)$.

(1.4) Remark: Notice that when $L_t \equiv 1/2\Delta$ (i.e. $a \equiv I$ and

$b \equiv 0$), $P(s,x;t,dy) = g(t-s,y-x)dy$ where g is the Gauss kernel given in I.4.1.

Throughout the rest of this section we will be working with the situation described in Theorem (1.2).

We first observe that when $\varphi \in C_0^\infty(\mathbb{R}^N)$, $u_{T,\varphi}$ is the unique $u \in C_b^{1,2}([0,T]\times\mathbb{R}^N)$ such that

$$\begin{aligned}&(\partial_s + L_s)u = 0 \text{ in } [0,T)\times\mathbb{R}^N\\ &\lim_{s\uparrow T} u(s,.) = \varphi.\end{aligned} \tag{1.5}$$

The uniqueness follows from (1.3) upon taking $f = u$. To prove that $u = u_{T,\varphi}$ satisfies (1.5), note that

$$\begin{aligned}\partial_s u(s,x) &= \lim_{h\downarrow 0} (u(s+h,x) - u(s,x))/h \\ &= \lim_{h\downarrow 0} \left[u(s+h,x) - \int u(s+h,y)P(s,x;s+h,dy)\right]/h \\ &= -\lim_{h\downarrow 0} 1/h\int_s^{s+h} dt\int [L_t u(s+h,.)](y)P(s,x;t,dy) \\ &= -L_s u(s,x),\end{aligned}$$

where we have used the Chapman-Kolmogorov equation ((I.3.4)), followed by (1.3). We next prove an important estimate for the tail distribution of the measure $P(s,x;t,\bullet)$.

(1.6) <u>Lemma</u>: Let $A = \sup_{t,x} \|a(t,x)\|_{op}$ and $B = \sup_{t,x} |b(t,x)|$. Then for all $0 \le s < T$ and $R > N^{1/2}B(T-s)$:

$$P(s,x;T,B(x,R)^c) \le 2N\exp\left[-(R-N^{1/2}B)^2/2NA(T-s)\right]. \tag{1.7}$$

In particular, for each $T > 0$ and $q \in [1,\infty)$, there is a $C(T,q) < \infty$, depending only on N, A and B, such that

$$\int |y-x|^q P(s,x;t,dy) \le C(T,q)(t-s)^{q/2}, \quad 0 \le s < t \le T. \tag{1.8}$$

<u>Proof</u>: Let A and B be any pair of numbers which are strictly larger than the ones specified, and let $T > 0$ and $x \in \mathbb{R}^N$ be given. Choose $\eta \in C^\infty(\mathbb{R}^1)$ so that $0 \le \eta \le 1$, $\eta \equiv 1$ on $[-1,1]$, and $\eta \equiv 0$ off of $(-2,2)$. Given $M \ge 1$, define Φ_M: $\mathbb{R}^N \longrightarrow \mathbb{R}^N$ by

$$(\Phi_M(y))^i = \int_0^{y^i - x^i} \eta(\xi/M)d\xi, \quad 1 \le i \le N;$$

and consider the function

$$f_{M,\theta}(t,y) \equiv \exp\left[(\theta,\Phi_M(y))_{R^N} + (A|\theta|^2 + B|\theta|)(T-t)\right]$$

for $\theta \in \mathbb{R}^N$ and $(t,y) \in [0,T]x\mathbb{R}^N$. Clearly $f_{M,\theta} \in C_b^\infty([0,T]x\mathbb{R}^N)$. Moreover, for sufficiently large M's, $(\partial_t + L_t)f_{M,\theta} \le 0$. Thus, by (1.3):

$$\int f_{M,\theta}(T,y)P(s,x;T,dy) \le f_{M,\theta}(s,x)$$

for all sufficiently large M's. After letting $M \longrightarrow \infty$ and applying Fatou's lemma, we now get:

$$\int \exp[(\theta,y-x)_{\mathbb{R}^N}]P(s,x;T,dy) \le \exp[(A|\theta|^2 + B|\theta|)(T-s)]. \tag{1.9}$$

Since (1.9) holds for all choices of A and B strictly larger than those specified, it must also hold for the ones which we were given.

To complete the proof of (1.7), note that:

$$P(s,x;T,B(x,R)^c) \le \sum_{i=1}^{N} P(s,x;T,\{y: |y^i - x^i| \ge R/N^{1/2}\})$$
$$\le 2N \max_{\theta \in S^{N-1}} P(s,x;T,\{y: (\theta,y-x)_{\mathbb{R}^N} \ge R/N^{1/2}\})$$

and, by (1.9)

$$P(s,x;T,\{y: (\theta,y-x)_{\mathbb{R}^N} \ge R/N^{1/2}\})$$

$$\le e^{-\lambda R/N^{1/2}}\int \exp[\lambda(\theta,y-x)_{\mathbb{R}^N}]P(s,x;T,dy)$$

$$\le \exp[\lambda^2 A(T-s)/2 - \lambda(R-N^{1/2}B(T-s))/N^{1/2}]$$

for all $\theta \in S^{N-1}$ and $\lambda > 0$. Hence, if $R > N^{1/2}B(T-s)$ and we take $\lambda = (R-N^{1/2}B(T-s))/N^{1/2}$, we arrive at (1.7). Q.E.D.

In view of (1.8), it is now clear from exercise (I.3.8) that for each $s \ge 0$ and each initial distribution $\mu_0 \in \mathbf{M}_1(\mathbb{R}^N)$ there is a unique $P_{s,\mu_0} \in \mathbf{M}_1(\Omega)$ such that

$$\begin{aligned} P_{s,\mu_0}&(x(t_0)\in\Gamma_0,\dots,x(t_n)\in\Gamma_n) \\ &=\int_{\Gamma_0}\mu_0(dx_0)\int_{\Gamma_1}P(s+t_0,x_0;s+t_1,dx_1) \\ &\qquad\cdots\int_{\Gamma_n}P(s+t_{n-1},x_{n-1};s+t_n,dx_n) \end{aligned} \tag{1.10}$$

for all $n \ge 0$, $0 = t_0 < \cdots < t_n$, and $\Gamma_0,\dots,\Gamma_n \in \mathcal{B}_{\mathbb{R}^N}$. We will use the notation $\underline{P}_{s,x}$ in place of P_{s,δ_x}.

(1.11) <u>Theorem</u>: The map $(s,x) \in [0,\infty)\times\mathbb{R}^N \longmapsto P_{s,x} \in \mathbf{M}_1(\Omega)$ is continuous and, for each $\mu_0 \in \mathbf{M}_1(R^N)$, $P_{s,\mu_0} = \int P_{s,x}\mu_0(dx)$. Moreover, $P_{s,x}$ is the one and only $P \in \mathbf{M}_1(\Omega)$ which satisfies:

$$\begin{aligned} &P(x(0) = x) = 1 \\ &P(x(t_2) \in \Gamma \mid \mathcal{M}_{t_1}) = P(s+t_1,x(s+t_1);s+t_2,\Gamma)\ \text{(a.s.,}P) \end{aligned} \tag{1.12}$$

for all $0 \le t_1 < t_2$ and $\Gamma \in \mathcal{B}_{\mathbb{R}^N}$. Finally, if $t \ge 0$ and $\omega \longmapsto (P^t_{s,x})_\omega$ is a r.c.p.d. of $P_{s,x}|\mathcal{M}_t$, then $(P^t_{s,x})_\omega \circ \theta_t^{-1} = P_{s+t,x(t,\omega)}$ for $P_{s,x}$-almost every ω.

<u>Proof</u>: First observe that, by the last part of Theorem (1.2), $(s,x) \in [0,T]\times\mathbb{R}^N \longmapsto \int\varphi(y)P(s,x;T,dy)$ is bounded and

continuous for all $\varphi \in C_0^\infty(\mathbb{R}^N)$. Combining this with (1.7), one sees that this continues to be true for all $\varphi \in C_b(\mathbb{R}^N)$. Hence, by (1.10), for all $n \geq 1$, $0 < t_1 < \cdots < t_n$, and $\varphi_1, \ldots, \varphi_n \in C_b(\mathbb{R}^N)$, $E^{P_{s,x}}[\varphi_1(x(t_1))\cdots\varphi_n(x(t_n))]$ is a bounded continuous function of $(s,x) \in [0,\infty)\times\mathbb{R}^N$. Now suppose that $(s_k,x_k)\longrightarrow(s,x)$ in $[0,\infty)\times\mathbb{R}^N$ and observe that, by (1.8) and (I.2.15), the sequence $\{P_{s_k,x_k}\}$ is relatively compact in $\mathbf{M}_1(\Omega)$. Moreover, if $\{P_{s_{k'},x_{k'}}\}$ is a convergent subsequence and P is its limit, then:

$$
\begin{aligned}
&E^P[\varphi_1(x(t_1))\cdots\varphi_n(x(t_n))] \\
&\quad = \lim_{k'\to\infty} E^{P_{s_{k'},x_{k'}}}[\varphi_1(x(t_1))\cdots\varphi_n(x(t_n))] \\
&\quad = E^{P_{s,x}}[\varphi_1(x(t_1))\cdots\varphi_n(x(t_n))]
\end{aligned}
$$

for all $n \geq 1$, $0 < t_1 < \cdots < t_n$, and $\varphi_1, \ldots, \varphi_n \in C_b(\mathbb{R}^N)$. Hence, $P = P_{s,x}$, and so we conclude that $P_{s_k,x_k}\longrightarrow P_{s,x}$. The fact that $P_{s,\mu_0} = \int P_{s,x}\mu_0(dx)$ is elementary now that we know that $(s,x)\longmapsto P_{s,x}$ is measurable.

Our next step is to prove the final assertion concerning $(P^t_{s,x})_\omega$. When $t = 0$, there is nothing to do. Assume that $t > 0$. Given $m,n \in Z^+$, $0 < \sigma_1 < \cdots < \sigma_m < t$, $0 < \tau_1 < \cdots < \tau_n$, and $\Delta_1, \ldots, \Delta_m, \Gamma_1, \ldots, \Gamma_n \in \mathcal{B}_{\mathbb{R}^N}$, set $A = \{x(\sigma_1)\in\Delta_1, \ldots, x(\sigma_m)\in\Delta_m\}$ and $B = \{x(\tau_1)\in\Gamma_1, \ldots, x(\tau_n)\in\Gamma_n\}$. Then:

$$
\int_A P_{s+t,x(t,\omega)}(B)P_{s,x}(d\omega)
$$

$$= \int_{\Delta_1} P(s,x;s+\sigma_1,dx_1)$$

$$\cdots\int_{\Delta_m} P(s+\sigma_{m-1},x_{m-1};s+\sigma_m,dx_m)$$

$$\times\int_{\mathbb{R}^N} P(s+\sigma_m,x_m;s+t,dy_0)\times\int_{\Gamma_1} P(s+t,y_0;s+t+\tau_1,dy_1)$$

$$\cdots\int_{\Gamma_n} P(s+t+\tau_{n-1},y_{n-1};s+t+\tau_n,dy_n)$$

$$= P_{s,x}(x(\sigma_1)\in\Delta_1,\ldots,x(\sigma_m)\in\Delta_m,x(t+\tau_1)\in\Gamma_1,\ldots,x(t+\tau_n)\in\Gamma_n)$$

$$= \int_A (P^t_{s,x})_\omega\circ\theta_t^{-1}(B)P_{s,x}(d\omega).$$

Hence, for all $A \in \mathcal{M}_t$ and $B \in \mathcal{M}$:

$$\int_A (P^t_{s,x})_\omega\circ\theta_t^{-1}(B)P_{s,x}(d\omega) = \int_A P_{s+t,x(t,\omega)}(B)P_{s,x}(d\omega).$$

Therefore, for each $B \in \mathcal{M}$, $(P^t_{s,x})_\omega\circ\theta_t^{-1}(B) = P_{s+t,x(t,\omega)}(B)$ (a.s.,$P_{s,x}$). Since $\mathcal{M}$ is countably generated, this, in turn, implies that $(P^t_{s,x})_\omega\circ\theta_t^{-1} = P_{s+t,x(t,\omega)}$ (a.s.,$P_{s,x}$).

Finally, we must show that $P_{s,x}$ is characterized by (1.12). That $P_{s,x}$ satisfies (1.12) is a special case of the result proved in the preceding paragraph. On the other hand, if $P \in M_1(\Omega)$ satisfies (1.12), then one can easily work by induction on $n \geq 0$ to prove that P satisfies (1.10) with $\mu_0 = \delta_x$.

Q.E.D.

(1.13) <u>Corollary</u>: For each $(s,x) \in [0,\infty)\times\mathbb{R}^N$, $P_{s,x}$ is the unique $P \in \mathbf{M}_1(\Omega)$ which satisfies $P(x(0) = x) = 1$ and:

$$E^P[\varphi(x(t_2)) - \varphi(x(t_1))|\mathcal{M}_{t_1}]$$

$$= E^P\left[\int_{t_1}^{t_2}[L_{s+t}\varphi](x(t))dt\,|\mathcal{M}_{t_1}\right] \text{ (a.s.,P)} \tag{1.14}$$

for all $0 \leq t_1 \leq t_2$ and $\varphi \in C_0^\infty(\mathbb{R}^N)$.

Proof: To see that $P_{s,x}$ satisfies (1.14), note that, by (1.12) and (1.3):

$$E^P[\varphi(x(t_2)) - \varphi(x(t_1))|\mathcal{M}_{t_1}]$$
$$= \int \varphi(y)P(s+t_1,x(s+t_1);s+t_2,dy) - \varphi(x(t_1))$$
$$= \int_{t_1}^{t_2} dt \int [L_{s+t}\varphi](y)P(s+t_1,x(s+t_1);s+t,dy)$$
$$= \int_{t_1}^{t_2} E^{P_{s,x}}[[L_{s+t}\varphi](x(t))|\mathcal{M}_{t_1}]dt$$
$$= E^{P_{s,x}}\left[\int_{t_1}^{t_2}[L_{s+t}\varphi](x(t))dt|\mathcal{M}_{t_1}\right] \text{ (a.s.,} P_{s,x}).$$

Conversely, if P satisfies (1.14) then it is easy to check that

$$E^P[f(t_2,x(t_2)) - f(t_1,x(t_1))|\mathcal{M}_{t_1}]$$
$$= E^P\left[\int_{t_1}^{t_2}[(\partial_t + L_{s+t})f(t,x(t))dt|\mathcal{M}_{t_1}\right] \text{ (a.s.,P)} \tag{1.15}$$

for all $0 \le t_1 < t_2$ and $f \in C_b^{1,2}([0,\infty)\times\mathbb{R}^N)$. In particular, if $\varphi \in C_0^\infty(\mathbb{R}^N)$ and $u(t,y) = \int\varphi(\eta)P(s+t,y;s+t_2,d\eta)$, then, by the last part of Theorem (1.2) together with (1.5), $u \in C_b^{1,2}([0,\infty)\times\mathbb{R}^N)$, $(\partial_t + L_{s+t})u = 0$ for $t \in [0,t_2)$, and $u(t_2,\cdot)$ $= \varphi$. Hence, from (1.15) with $f = u$:

$$E^{P_{s,x}}[\varphi(x(t_2))|\mathcal{M}_{t_1}] = u(t_1,x(t_1)) \text{ (a.s.,P)}.$$

Combined with $P(x(0) = x) = 1$, this proves that P satisfies the condition in (1.12) characterizing $P_{s,x}$. Q.E.D.

(1.16) Remark: The characterization of $P_{s,x}$ given in Corollary (1.13) has the great advantage that it only involves L_t and does not make direct reference to $P(s,x;t,\cdot)$.

Since, in most situations, L_t is a much more primitive quantity than the associated quantity $P(s,x;t,\cdot)$, it should be clear that there is considerable advantage to having $P_{s,x}$ characterized directly in terms of L_t itself. In addition, (1.14) has great intuitive appeal. What it says is that, in some sense, $P_{s,x}$ sees the paths ω as the "integral curves of L_t initiating from x at time s." Indeed, (1.14) can be converted into the statement that

$$E^P[\varphi(x(t+h)) - \varphi(x(t))|\mathcal{M}_t] = hL_t\varphi(x(t)) + o(h),\ h\downarrow 0,$$

which, in words, says that "based on complete knowledge of the past up until time t, the best prediction about the P-value of $\varphi(x(t+h)) - \varphi(x(t))$ is, up to lower order terms in h, $hL_t\varphi(x(t))$." This intuitive idea is expanded upon in the following exercise.

(1.17) Exercise: Assume that $a \equiv 0$ and that b is independent of t. Show, directly from (1.14) that in this case $P_{0,x} = \delta_{X(\cdot,x)}$ where $X(\cdot,x)$ is the integral curve of the vector field b starting at x. In fact, you can conclude this fact about $P_{0,x}$ from $P(x(0) = x) = 1$ and the unconditional version of (1.14):

$$E^P[\varphi(x(t_2)) - \varphi(x(t_1))] = E^P\left[\int_{t_1}^{t_2}[L_{s+t}\varphi](x(t))dt\right]. \quad (1.14')$$

Finally, when $L_t \equiv 1/2\Delta$, show that the unconditional statement is not sufficient to characterize $\mathcal{W}_x$. (Hint: let X be an $\mathbb{R}^N$-valued Gaussian random variable with mean 0 and covariance I, denote by $P \in \mathbf{M}_1(\Omega)$ the distribution of the paths $t \longmapsto t^{1/2}X$, and check that (1.14') holds with this P but that $P \neq \mathcal{W}$.)

2. The Elements of Martingale Theory:

Let $P_{s,x}$ be as in section 1). Then (1.14) can be re-arranged to be the statement that:

$$E^{P_{s,x}}[X_\varphi(t_2)|\mathcal{M}_{t_1}] = X_\varphi(t_1) \ (\text{a.s.},P_{s,x}),\ 0 \le t_1 < t_2,$$

where

$$X_\varphi(t) = \varphi(x(t)) - \varphi(x(0)) - \int_0^t [L_{s+\tau}\varphi](x(\tau))d\tau. \qquad (2.1)$$

Loosely speaking, (2.1) is the statement that $t \longmapsto X_\varphi(t)$ is "conditionally constant" under $P_{s,x}$ in the sense that $X_\varphi(t_1)$ is "the best prediction about the $P_{s,x}$-value of $X_\varphi(t_2)$ given perfect knowledge of the past up to time t_1" (cf. remark (1.16)). Of course, this is another way of viewing the idea that $P_{s,x}$ sees the path ω as an "integral curve of L_t." Indeed, if ω were "truly an integral curve of L_t", we would have that $X_\varphi(\cdot,\omega)$ is "truly constant." The point is that we must settle for $X_\varphi(\cdot,\omega)$ being constant only in the sense that it is "predicted to be constant." Since these conditionally constant processes arise a great deal and have many interesting properties, we will devote this section to explaining a few of the basic facts about them.

Let $(E,\mathcal{F},P)$ be a probability space and $\{\mathcal{F}_t: t \ge 0\}$ a non-decreasing family of sub-σ-algebras of $\mathcal{F}$. A map X on $[0,\infty)\times E$ into a measurable space is said to be ($\{\mathcal{F}_t\}$-)<u>progressively measurable</u> if its restriction to $[0,T]\times E$ is $\mathcal{B}_{[0,T]}\times\mathcal{F}_T$-measurable for each $T \ge 0$. A map X on $[0,\infty)\times E$ with values in a topological space is said to be, respectively, <u>right continuous</u> (<u>P-a.s. right continuous</u>) or

<u>continuous</u> (<u>P-a.s. continuous</u>) if for every (P-almost every) $\xi \in E$ the map $t \longmapsto X(t,\xi)$ is right continuous or continuous.

(2.2) <u>Exercise</u>: Show that the notion of progressively measurable coincides with the notion of measurability with respect to the σ-algebra of <u>progressively measurable sets</u> (i.e. those subsets Γ of $[0,\infty)\times E$ for which χ_Γ is a progressively measurable function). In addition, show that if X is $\{\mathcal{F}_t\}$-<u>adapted</u> in the sense that $X(t,.)$ is $\mathcal{F}_t$-measurable for each $t \geq 0$, then X is $\{\mathcal{F}_t\}$-progressively measurable if it is right continuous.

A $\mathbb{C}$-valued map X on $[0,\infty)\times E$ is called a <u>martingale</u> if X is a right-continuous, $\{\mathcal{F}_t\}$-progressively measurable function such that $X(t) \in L^1(P)$ for all $t \geq 0$ and

$$X(t_1) = E^P[X(t_2)|\mathcal{F}_{t_1}] \text{ (a.s.,P)}, \quad 0 \leq t_1 < t_2. \tag{2.3}$$

Unless it is stated otherwise, it should be assumed that all martingales are real-vauled. An $\mathbb{R}^1$-valued map X on $[0,\infty)\times E$ is said to be a <u>sub-martingale</u> if X is a right continuous, $\{\mathcal{F}_t\}$-progressivly measurable function such that $X(t) \in L^1(P)$ for every $t \geq 0$ and

$$X(t_1) \leq E^P[X(t_2)|\mathcal{F}_{t_1}] \text{ (a.s.,P)}, \quad 0 \leq t_1 < t_2. \tag{2.4}$$

We will often summarize these statements by saying that <u>$(X(t),\mathcal{F}_t,P)$ is a martingale</u> (<u>sub-martingale</u>).

(2.5) <u>Example</u>: Besides the source of examples provided by (2.1), a natural way in which martingales arrise is the following. Let $X \in L^1(P)$ and define $X(t) = E^P[X|\mathcal{F}_{[t]}]$ (we use $[r]$ to denote the integer part of an $r \in \mathbb{R}^1$). Then it is an easy matter to check that $(X(t),\mathcal{F}_t,P)$ is a martingale. More generally, let Q be a totally finite measure on $(E,\mathcal{F})$ and assume that $Q\restriction\mathcal{F}_t << P\restriction\mathcal{F}_t$ for each $t \geq 0$. Then $(X(t),\mathcal{F}_t,P)$ is a martingale when $X(t)$ denotes the Radon-Nikodym derivative of $Q\restriction\mathcal{F}_{[t]}$ with respect to $P\restriction\mathcal{F}_{[t]}$. It turns out that these examples generate, in a resonable sense, all the examples of martingales.

The following statement is an easy consequence of Jensen's inequality.

(2.7) <u>Lemma</u>: Let $(X(t),\mathcal{F}_t,P)$ be a martingale (sub-martingale) with values in the closed interval I. Let φ be a continuous function on I which is convex (and non-decreasing). If $\varphi\circ X(t) \in L^1(P)$ for every $t \geq 0$, then $(\varphi\circ X(t),\mathcal{F}_t,P)$ is a sub-martingale. In particular, if $q \in [1,\infty)$ and $(X(t),\mathcal{F}_t,P)$ is an L^q-martingale (non-negative L^q-sub-martingale) (i.e. $X(t) \in L^q(P)$ for all $t \geq 0$), then $(|X(t)|^q,\mathcal{F}_t,P)$ is a sub-martingale.

(2.8) <u>Theorem</u> (Doob's Inequality): Let $(X(t),\mathcal{F}_t,P)$ be a non-negative sub-martingale. Then, for each $T > 0$:

$$P(X^*(T)\geq R) \leq E^P[X(T),\ X^*(T)\geq R],\ R \geq 0, \tag{2.9}$$

where

$$X^*(T) \equiv \sup_{0\leq t\leq T}|X(t)|,\ T \geq 0. \tag{2.10}$$

In particular, for each $T > 0$, the family $\{X(t): t \in [0,T]\}$ is uniformly P-integrable; and so, for every $s \geq 0$, $X(t) \longrightarrow X(s)$ in $L^1(P)$ as $t \downarrow s$. Finally, if $X(T) \in L^q(P)$ for some $T > 0$ and $q \in (1,\infty)$, then

$$E^P[X^*(T)^q]^{1/q} \leq q/(q-1)E^P[X(T)^q]^{1/q}. \tag{2.11}$$

Proof: Given $n \geq 1$, note that:

$$P(\max_{0\leq k\leq n} X(kT/n)\geq R) = \sum_{\ell=0}^{n} P(X(\ell T/n)\geq R \;\&\; \max_{0\leq k<\ell} X(kT/n)<R)$$

$$\leq 1/R \sum_{\ell=0}^{n} E^P[X(\ell T/n),\ X(\ell T/n)\geq R \;\&\; \max_{0\leq k<\ell} X(kT/n)<R]$$

$$\leq 1/R \sum_{\ell=0}^{n} E^P[X(T),\ X(\ell T/n)\geq R \;\&\; \max_{0\leq k<\ell} X(kT/n)<R]$$

$$\leq \frac{1}{R}E^P[X(T),\ X^*(T)\geq R].$$

Since $\max_{0\leq k\leq n} X(kT/n) \longrightarrow X^*(T)$ as $n \longrightarrow \infty$, the proof of (2.9) is complete.

To prove the uniform P-integrability statement, note that for $t \in [0,T]$:

$$E^P[X(t),\ X(t)\geq R] \leq E^P[X(T),\ X(t)\geq R] \leq E^P[X(T),\ X^*(T)\geq R].$$

Since $X(T) \in L^1(P)$ and $P(X^*(T)\geq R) \longrightarrow 0$ as $R \longrightarrow \infty$, we see that:

$$\lim_{R\to\infty} \sup_{t\in[0,T]} E^P[X(t),\ X(t)\geq R] = 0.$$

Finally, to prove (2.11), we show that for any pair of non-negative random variables X and Y satisfying $P(Y\geq R) \leq E^P[X, Y\geq R]$, $R \geq 0$, $\| Y \|_{L^q(P)} \leq q/(q-1) \| X \|_{L^q(P)}$, $q \in (1,\infty)$. Clearly, we may assume ahead of time that Y is bounded. The proof then is a simple integration by parts:

$$E^P[Y^q] = q\int_0^\infty R^{q-1}P(Y\geq R)dR \leq q\int_0^\infty R^{q-2}E^P[X,\ Y\geq R]dR$$

$$= q\int_0^\infty R^{q-2}dR\int_0^\infty P(X\geq r,\ Y\geq R)dr = q/(q-1)\int_0^\infty E[Y^{q-1},\ X\geq r]dr$$

$$= q/(q-1)E^P[Y^{q-1}X] \leq q/(q-1)E^P[Y^q]^{(q-1)/q}E^P[X^q]^{1/q}.$$

Q.E.D.

A function τ: $E\longrightarrow[0,\infty]$ is said to be a $(\{\mathcal{F}_t\}$-)<u>stopping time</u> if $\{\tau \leq t\} \in \mathcal{F}_t$ for every $t \geq 0$. Given a stopping time τ, define $\mathcal{F}_\tau$ to be the collection of sets $\Gamma \subseteq E$ such that $\Gamma\cap\{\tau \leq t\} \in \mathcal{F}_t$ for all $t \geq 0$.

(2.12) <u>Exercise</u>: In the following, σ and τ are stopping times and X is a progressively measurable function. Prove each of the statements:

i) $\mathcal{F}_\tau$ is a sub-σ-algebra of $\mathcal{F}$, $\mathcal{F}_\tau = \mathcal{F}_T$ if $\tau \equiv T$, and τ is $\mathcal{F}_\tau$-measurable;

ii) $\xi \in E\longmapsto X(\tau,\xi) \equiv X(\tau(\xi),\xi)$ is $\mathcal{F}_\tau$-measurable;

iii) $\sigma + \tau$, $\sigma\vee\tau$, and $\sigma\wedge\tau$ are stopping times;

iv) if $\Gamma \in \mathcal{F}_\sigma$, then $\Gamma\cap\{\sigma \leq \tau\}$ and $\Gamma\cap\{\sigma < \tau\}$ are in $\mathcal{F}_{\sigma\wedge\tau}$;

v) if $\sigma \leq \tau$, then $\mathcal{F}_\sigma \subseteq \mathcal{F}_\tau$.

If you get stuck, see 1.2.4 in [S.&V.].

(2.13) <u>Theorem</u> (Hunt): Let $(X(t),\mathcal{F}_t,P)$ be a martingale (non-negative sub-martingale). Given stopping times σ and τ which satisfy $\sigma \leq \tau \leq T$ for some $T > 0$, $X(\sigma) = E^P[X(\tau)|\mathcal{F}_\sigma]$ ($X(\sigma) \leq E^P[X(\tau)|\mathcal{F}_\sigma]$) (a.s.,P). In particular, if $(X(t),\mathcal{F}_t,P)$

is a non-negative sub-martingale, then, for any $T > 0$, $\{X(\tau): \tau$ is a stopping time $\leq T\}$ is uniformly P-integrable.

<u>Proof</u>: Let $\mathcal{G}$ denote the set of all stopping times $\sigma: E \longrightarrow [0,T]$ such that $X(\sigma) = E^P[X(T)|\mathcal{F}_\sigma]$ $(X(\sigma) \leq E^P[X(T)|\mathcal{F}_\sigma])$ (a.s.,P) for every martingale (non-negative sub-martingale) $(X(t),\mathcal{F}_t,P)$. Then, for any non-negative sub-martingale $(X(t),\mathcal{F}_t,P)$:

$$\lim_{R\to\infty} \sup_{\sigma\in\mathcal{G}} E^P[X(\sigma),\ X(\sigma)\geq R] \leq \lim_{R\to\infty} \sup_{\sigma\in\mathcal{G}} E^P[X(T),\ X(\sigma)\geq R]$$

$$\leq \lim_{R\to\infty} E^P[X(T),\ X^*(T)\geq R] = 0;$$

and so $\{X(\sigma): \sigma \in \mathcal{G}\}$ is uniformly P-integrable. In particular, if σ is a stopping time which is the non-increasing limit of elements of $\mathcal{G}$, then $\sigma \in \mathcal{G}$.

We next show that if σ is a stopping time which takes on only a finite number of values $0 = t_0 < \cdots < t_n = T$, then $\sigma \in \mathcal{G}$. To this end, let $\Gamma \in \mathcal{F}_\sigma$ be given and set $\Gamma_k = \Gamma \cap \{\sigma = t_k\}$. Then $\Gamma_k \in \mathcal{F}_{t_k}$ and so:

$$E^P[X(\sigma),\ \Gamma] = \sum_{k=0}^{n} E^P[X(t_k),\ \Gamma_k]$$

$$\overset{(\leq)}{=} \sum_{k=0}^{n} E^P[X(T),\ \Gamma_k] = E^P[X(T),\ \Gamma].$$

Now let $\sigma: E \longrightarrow [0,T]$ be any stopping time, and, for $n \geq 0$, set $\sigma_n = (([2^n\sigma]+1)/2^n)\wedge T$. By the preceding, $\sigma_n \in \mathcal{G}$ for each $n \geq 1$. In addition, $\sigma_n \downarrow \sigma$. Hence, we now know that every stopping time bounded by T is an element of $\mathcal{G}$. In particular, if $(X(t),\mathcal{F}_t,P)$ is a non-negative sub-martingale, then the set of $X(\sigma)$ as σ runs over steping times bounded by

T is uniformly P-integrable. Also, if $\sigma \le \tau \le T$ are stopping times, then for any martingale $(X(t),\mathcal{F}_t,P)$ we have:

$$E^P[X(\tau)|\mathcal{F}_\sigma] = E^P[E^P[X(T)|\mathcal{F}_\tau]|\mathcal{F}_\sigma] = E^P[X(T)|\mathcal{F}_\sigma] = X(\sigma)$$

(a.s.,P).

It remains to show that if $(X(t),\mathcal{F}_t,P)$ is a non-negative sub-martingale and $\sigma \le \tau \le T$ are stoping times, then $E^P[X(\tau)|\mathcal{F}_\sigma] \ge X(\sigma)$ (a.s.,P). Notice that, by the uniform integrability property already proved, we need only do this when σ and τ take values in a finite set $0 = t_0 < \cdots < t_n = T$. To handle this case, define

$$A(t) = \begin{cases} \sum_{k=0}^{\ell-1} E^P[X(t_{k+1})-X(t_k)|\mathcal{F}_{t_k}], & t\in[t_\ell,t_{\ell+1}) \text{ and } 0\le\ell<n \\ \sum_{k=0}^{n-1} E^P[X(t_{k+1})-X(t_k)|\mathcal{F}_{t_k}], & t\ge T. \end{cases}$$

Then, $t\longmapsto A(t)$ is P-almost surely non-decreasing and $(M(t),\mathcal{F}_t,P)$ is a martingale, where

$$M(t) = \begin{cases} X(t_\ell)-A(t), & t\in[t_\ell,t_{\ell+1}) \text{ and } 0\le\ell<n \\ X(T)-A(t), & t\ge T. \end{cases}$$

Hence:

$$\begin{aligned} E^P[X(\tau)|\mathcal{F}_\sigma] &= E^P[M(\tau)+A(\tau)|\mathcal{F}_\sigma] \\ &= X(\sigma) + E^P[A(\tau)-A(\sigma)|\mathcal{F}_\sigma] \ge X(\sigma) \text{ (a.s.,P)} \end{aligned}$$

Q.E.D.

(2.14) <u>Corollary</u>(Doob's Stopping Time Theorem): If $(X(t),\mathcal{F}_t,P)$ is a martingale (non-negative sub-martingale) and τ is a stopping time, then $(X(t\wedge\tau),\mathcal{F}_t,P)$ is a martingale (sub-martingale).

<u>Proof</u>: Let $0 \le s \le t$ and $\Gamma \in \mathcal{F}_s$. Then, since $\Gamma\cap\{\tau>s\} \in \mathcal{F}$:

$$E^P[X(t\wedge\tau),\ \Gamma] = E^P[X(t\wedge\tau),\ \Gamma\cap\{\tau>s\}] + E^P[X(t\wedge\tau),\ \Gamma\cap\{\tau\le s\}]$$
$$\overset{(\ge)}{=} E^P[X(s\wedge\tau),\ \Gamma\cap\{\tau>s\}] + E^P[X(s\wedge\tau),\ \Gamma\cap\{\tau\le s\}]$$
$$= E^P[X(s\wedge\tau),\ \Gamma].$$

Q.E.D.

(2.15) Exercise: Let φ: $[0,\infty)\longrightarrow\mathbb{R}^1$ be a right continuous function. Given $a < b$ and $T \in (0,\infty]$ we say that φ upcrosses [a,b] at least n times during [0,T) if there exist $0 \le s_1 < t_1 < \cdots < s_n < t_n < T$ such that $\varphi(s_m) < a$ and $\varphi(t_m) > b$ for each $1 \le m \le n$. Define $U(a,b;T) = \inf\{n \ge 0\colon \varphi$ does not upcross $[a,b]$ at least $(n+1)$ times during $[0,T)\}$ to be the number of times that φ upcrosses [a,b] during [0,T). Show that φ has a left limit (in $[-\infty,\infty]$) at every $t \in [0,T]$ if an only if $U(a,b;T) < \infty$ for all rational $a < b$. Also, check that if $U_m(a,b;T)$ is defined relative to the function $t\longmapsto \varphi(([2^m t]/2^m)$, then $U_m(a,b;T)\uparrow U(a,b;T)$ as $m\longrightarrow\infty$.

(2.16) Theorem (Doob's Upcrossing Inequality): Let $(X(t),\mathcal{F}_t,P)$ be a sub-martingale; and, for $a < b$, $\xi \in E$, and $T \in (0,\infty]$, define $U(a,b;T)(\xi)$ to be the number of times that $t\longmapsto X(t,\xi)$ upcrosses [a,b] during [0,T). Then:

(2.17) $\quad E^P[U(a,b;T)] \le E^P[(X(T)-a)^+]/(b-a),\ T \in (0,\infty).$

In particular, for P-almost all $\xi \in E$, $X(\cdot,\xi)$ has a left limit (in $[-\infty,\infty)$) at each $t \in (0,\infty)$. In addition, if $\sup_{T>0} E^P[X(T)^+] < \infty$ ($\sup_{T>0}E^P[|X(T)|] < \infty$), then $\lim_{t\to\infty}X(t)$ exists in $[-\infty,\infty)$ ($(-\infty,\infty)$) (a.s.,P).

Proof: In view of exercise (2.15), it suffices to prove that (2.17) holds with $U(a,b;T)$ replaced by $U_m(a,b;T)$ (cf.

the last part of (2.15)).

Given $m \geq 0$, set $X_m(t) = X(([2^m t]/2^m)$ and $\tau_0 \equiv 0$, and define σ_n and τ_n inductively for $n \geq 1$ by:

$$\sigma_n = (\inf\{t \geq \tau_{n-1}: X_m(t) < a\}) \wedge T$$

and

$$\tau_n = (\sup\{t \geq \sigma_n: X_m(t) > b\}) \wedge T.$$

Clearly the σ_n's and τ_n's are stopping times which are bounded by T, and $U_m(a,b;T) = \max\{n \geq 0: \tau_n < T\}$. Thus, if $Y_m(t) \equiv (X_m(t)-a)^+/(b-a)$, then:

$$U_m(a,b;T) \leq \sum_{n=0}^{[2^m T]} (Y_m(\tau_n) - Y_m(\sigma_n));$$

and so:

$$\begin{aligned} Y_m(T) &\geq Y_m(T) - Y_m(0) \\ &\geq U_m(a,b;T) + \sum_{n=0}^{[2^m T]} (Y_m(\sigma_{n+1}) - Y_m(\tau_n)). \end{aligned}$$

But $(Y_m(t),\mathcal{F}_t,P)$ is a non-negative sub-martingale and therefore:

$$E^P[Y_m(\sigma_{n+1}) - Y_m(\tau_n)] \geq 0.$$

At the same time, $((X(t)-a)^+,\mathcal{F}_t,P)$ is a sub-martingale and therefore

$$E^P[Y_m(T)] \leq E^P[(X(T)-a)^+]/(b-a).$$

Q.E.D.

(2.18) <u>Corollary</u>: If $(X(t),\mathcal{F}_t,P)$ is a P-almost surely continuous sub-martingale, then $\lim_{t\to\infty} X(t)$ exists (in $[-\infty,\infty)$) (a.s.,P) on the set $B \equiv \{\xi \in E: \sup_{t\geq 0} X(t,\xi) < \infty\}$. In the case of P-almost surely continuous martingales, the conclusion is that the limit exists P-almost surely in $(-\infty,\infty)$ on B.

Proof: Without loss in generality, we assume that $X(0) \equiv 0$. Given $R > 0$, set $\tau_R = \inf\{t \geq 0: \sup_{0 \leq s \leq t} X(s) \geq R\}$ and define $X_R(t) = X(t \wedge \tau_R)$. Then τ_R is a stopping time and $X_R(t) \leq R$, $t \geq 0$, (a.s.,P). Hence, $(X_R(t), \mathcal{F}_t, P)$ is a sub-martingale and $E^P[X_R(T)^+] \leq R$ for all $T \geq 0$. In particular, $\lim_{t \to \infty} X(t)$ exists (in $[-\infty,\infty)$) (a.s.,P) on $\{\tau_R = \infty\}$. Since this is true for every $R > 0$, we now have the desired conclusion in the sub-martingale case. The martingale case follows from this, the observation that $E^P[|X_R(T)|] = 2E^P[X_R(T)^+] - E^P[X_R(0)]$, and Fatou's lemma.

Q.E.D.

(2.19) Exercise: Prove each of the following statements.

i) $(X(t), \mathcal{F}_t, P)$ is a uniformly P-integrable martingale if and only if $X(\infty) = \lim_{t \to \infty} X(t)$ exists in $L^1(P)$, in which case $X(t) \longrightarrow X(\infty)$ (a.s.,P) and $X(\tau) = E^P[X(\infty)|\mathcal{F}_\tau]$ (a.s.,P) for each stopping time τ.

ii) If $q \in (1,\infty)$ and $(X(t), \mathcal{F}_t, P)$ is a martingale, then $(X(t), \mathcal{F}_t, P)$ is $L^q(P)$-bounded (i.e. $\sup_{t \geq 0} E^P[|X(t)|^q] < \infty$) if and only if $X(\infty) = \lim_{t \to \infty} X(t)$ in $L^q(P)$, in which case $X(t) \longrightarrow X(\infty)$ (a.s.,P) and $X(\tau) = E^P[X(\infty)|\mathcal{F}_\tau]$ (a.s.,P) for each stopping time τ.

iii) Suppose that $X: [0,\infty) \times E \longrightarrow \mathbb{R}^1$ ($[0,\infty)$) is a right continuous progressively measurable function and that $X(t) \in L^1(P)$ for each t in a dense subset D of $[0,\infty)$. If $X(s) \overset{(\leq)}{=} E^P[X(t)|\mathcal{F}_s]$ (a.s.,P) for all $s,t \in D$ with $s < t$, then

$(X(t),\mathcal{F}_t,P)$ is a martingale (non-negative sub-martingale).

iv) Let Ω be a Polish space, $P \in M_1(\Omega)$, and $\mathcal{A}$ a countably generated sub-σ-algebra of $\mathcal{B}_\Omega$. Then there exists a nested sequence $\{\Pi_n\}$ of finite partitions of Ω into $\mathcal{A}$-measurable sets such that $\bigcup_{n=1}^{\infty} \sigma(\Pi_n)$ generates $\mathcal{A}$. In addition, if $\omega \longmapsto P_\omega^n$ is defined by

$$P_\omega^n(B) = \sum_{A\in\Pi_n} [P(B\cap A)/P(A)]\chi_A(\omega)$$

for $B \in \mathcal{B}_\Omega$ ($0/0 \equiv 0$ here), then there is a P-null set $\Lambda \in \mathcal{A}$ such that $P_\omega^n \longrightarrow P_\omega$ in $M_1(\Omega)$ for each $\omega \notin \Lambda$. Finally, P_ω can be defined for $\omega \notin \Lambda$ so that $\omega \longmapsto P_\omega$ becomes a r.c.p.d. of $P|\mathcal{A}$.

(2.20) <u>Theorem</u>: Assume that $\mathcal{F}_t$ is countably generated for each $t \geq 0$. Let τ be a stopping time and suppose that $\omega \longmapsto P_\omega$ is a c.p.d. of $P|\mathcal{F}_\tau$. Let $X: [0,\infty)\times E \longrightarrow R^1$ be a right continuous progressively measurable function and assume that $X(t) \in L^1(P)$ for all $t \geq 0$. Then $(X(t),\mathcal{F}_t,P)$ is a martingale if and only if $(X(t\wedge\tau),\mathcal{F}_t,P)$ is a martingale and there is a P-null set $\Lambda \in \mathcal{F}_\tau$ such that $(X(t) - X(t\wedge\tau),\mathcal{F}_t,P_\omega)$ is a martingale for each $\omega \notin \Lambda$.

<u>Proof</u>: Set $Y(t) = X(t) - X(t\wedge\tau)$. Assuming that $(X(t\wedge\tau),\mathcal{F}_t,P)$ is a martingale and that $(X(t) - X(t\wedge\tau),\mathcal{F}_t,P_\omega)$ is a martingale for each ω outside of an $\mathcal{F}_\tau$-measurable P-null, we have, for each $s < t$ and $\Gamma \in \mathcal{F}_s$:

$$\begin{aligned} E^P[X(t)-X(s),\ \Gamma] &= E^P[Y(t)-Y(s),\ \Gamma] + E^P[X(t\wedge\tau)-X(s\wedge\tau),\ \Gamma] \\ &= \int E^{P_\omega}[Y(t)-Y(s),\ \Gamma]P(d\omega) = 0. \end{aligned}$$

That is, $(X(t),\mathcal{F}_t,P)$ is a martingale.

Next, assume that $(X(t),\mathcal{F}_t,P)$ is a martingale. Then $(X(t\wedge\tau),\mathcal{F}_t,P)$ is a martingale by Doob's stopping time theorem. To see that $(Y(t),\mathcal{F}_t,P_\omega)$ is a martingale for each ω outside of a P-null set $\Lambda \in \mathcal{F}_\tau$, we proceed as follows. Given $0 \leq s < t$, $\Gamma \in \mathcal{F}_s$, and $A \in \mathcal{F}_\tau$, we have:

$$\int_A E^{P_\omega}[Y(t),\ \Gamma]P(d\omega) = E^P[Y(t),\ \Gamma\cap A]$$

$$= E^P[Y(t),\ \Gamma\cap A\cap\{\tau\leq s\}] + E^P[Y(t),\ \Gamma\cap A\cap\{s<\tau\leq t\}].$$

Note that, by exercise (2.12),

$$\Gamma\cap A\cap\{s<\tau\leq t\} = (\Gamma\cap\{s<\tau\})\cap A\cap\{\tau\leq t\} \in \mathcal{F}_\tau.$$

Thus $E^P[Y(t),\ \Gamma\cap A\cap\{s<\tau\leq t\}] = E^P[X(\tau)-X(\tau),\ \Gamma\cap A\cap\{s<\tau\leq t\}] = 0$. At the same time, $\Gamma\cap A\cap\{\tau\leq s\} \in \mathcal{F}_s$ and so:

$$E^P[Y(t),\ \Gamma\cap A\cap\{\tau\leq s\}] = E^P[X(s)-X(\tau),\ \Gamma\cap A\cap\{\tau\leq s\}]$$

$$= E^P[Y(s),\ \Gamma\cap A] = \int_A E^{P_\omega}[Y(s),\ \Gamma]P(d\omega).$$

We have therfore proved that:

$$\int_A E^{P_\omega}[Y(t),\ \Gamma]P(d\omega) = \int_A E^{P_\omega}[Y(s),\ \Gamma]P(d\omega).$$

Since $\mathcal{F}_s$ is countably generated, we conclude from this that there is a P-null set $\Lambda \in \mathcal{F}_\tau$ such that for all $\omega \notin \Lambda$: $\{Y(t): t\in\mathbb{Q}\cap[0,\infty)\} \subseteq L^1(P_\omega)$ and $Y(s) = E^{P_\omega}[Y(t)|\mathcal{F}_s]$ (a.s.,P_ω) for all rational $s < t$. In view of the iii) in exercise (2.19), this completes the proof.

Q.E.D.

(2.21) <u>Exercise</u>: Let everything be as in Theorem (2.20), only this time assume that $X(t) \geq 0$ for all $t \geq 0$. Show that

$(X(t), \mathcal{F}_t, P)$ is a sub-martingale if and only if $(X(t\wedge\tau), \mathcal{F}_t, P)$ is a sub-martingale and there is a P-null set $\Lambda \in \mathcal{F}_\tau$ such that $(X(t)\chi_{[0,t]}(\tau), \mathcal{F}_t, P_\omega)$ is a sub-martingale for all $\omega \notin \Lambda$.

The rest of this section is devoted to a particularly useful special case of the reknowned Doob-Meyer decomposition theorem. What their theorem says is that, under mild technical hypotheses, every sub-martingale $(X(t), \mathcal{F}_t, P)$ is the sum of a martingale $(M(t), \mathcal{F}_t, P)$ and a right-continuous, progressively measurable function A: $[0,\infty)\times E \longrightarrow [0,\infty)$ having the property that $t \longmapsto A(t)$ is P-almost surely non-decreasing. Moreover, A can be chosen so that $A(0) = 0$ and $t \longmapsto A(t)$ is "nearly left-continuous" (precisely, A is a "$\{\mathcal{F}_t\}$-predictable"); and, among such non-decreasing processes, there is only one whose difference from X is a $\{\mathcal{F}_t\}$-martingale under P. We have already seen a special case of this decompostion in our proof of Theorem (2.13) (cf. the construction of the processes M and A made there). The idea used in the proof of Theorem (2.13) is due to Doob and it works whenever $t \longmapsto X(t)$ is piece-wise constant. What Meyer did is show that, in general, A can be realized as the limit of A's constructed for piece-wise constant approximations to X. Simple as this procedure may sound, it is frought with technical difficulties. To avoid these difficulties, and because we will not have great need for the general result, we will content ourselves with the special case of sub-martingales $(X(t)^2, \mathcal{F}_t, P)$ where $(X(t), \mathcal{F}_t, P)$ is a real-valued, P-almost surely continous, $L^2(P)$-martingale. Our proof of existence for this case is based on ideas of K. Itô. The uniqueness assertion will be a consequence of the following simple lemma.

(2.22) <u>Lemma</u>: Let $(X(t),\mathcal{F}_t,P)$ be a martingale and $A: [0,\infty)\times E\longrightarrow R^1$ a right-continuous progressively measurable function which is P-almost surely continuous and of local bounded variation (i.e. for each $T > 0$, the total variation $|A|(T,\xi)$ of $A(\cdot,\xi)\restriction[0,T]$ is finite for P-almost every $\xi \in E$). Then, assuming that

$$E^P[\sup_{0\leq t\leq T}|X(t)|(|A|(T)+|A(0)|)] < \infty$$

for all $T > 0$, $(X(t)A(t)-\int_0^t X(s)A(ds),\mathcal{F}_t,P)$ is a martingale.

<u>Proof</u>: Let $0 \leq s < t$ and $\Gamma \in \mathcal{F}_s$ be given. Then:

$$\begin{aligned} &E^P[X(t)A(t)-X(s)A(s),\Gamma] \\ &\quad = E^P\Big[\sum_{k=0}^{n-1}[X(u_{n,k+1})A(u_{n,k+1})-X(u_{n,k})A(u_{n,k})],\Gamma\Big] \\ &\quad = E^P\Big[\sum_{k=0}^{n-1}[X(u_{n,k+1})(A(u_{n,k+1})-A(u_{n,k}))],\Gamma\Big], \end{aligned}$$

where $u_{n,k} \equiv s + k(t - s)/n$. Since $u\longmapsto X(u)$ is right continuous and $u\longmapsto A(u)$ is P-almost surely continuous and of local bounded variation,

$$\sum_{k=1}^{n-1}[X(u_{n,k+1})(A(u_{n,k+1})-A(u_{n,k}))]\longrightarrow\int_0^t X(s)A(ds) \quad (a.s.,P);$$

and our integrablility assumption allows us to conclude that this convergence takes place in $L^1(P)$. Q.E.D.

(2.23) Theorem: Let $(X(t),\mathcal{F}_t,P)$ be a P-almost surely continuous martingale and define $\zeta = \sup\{t\geq 0: |X|(t) < \infty\}$. Then, P-almost surely, $X(t\wedge\zeta) = X(0)$, $t \geq 0$. In particular, if $P(X(t) = X(s)) = 0$ for all $0 \leq s < t$, then $t\longmapsto X(t)$ is P-almost never of bounded variation on any interval.

Proof: Without loss of generality, we assume that $X(0) \equiv 0$ and that $X(\cdot,\xi)$ is continuous for all $\xi \in E$.

For $R > 0$, define $\zeta_R = \sup\{t\geq 0: |X|(t) < R\}$. Then ζ_R is a stopping time for each $R > 0$ and $\zeta_R\longrightarrow\zeta$ as $R\longrightarrow\infty$. Moreover, by Lemma (2.22) with $X(\cdot)$ and $A(\cdot)$ replaced by $X(\cdot\wedge\zeta_R)$ and $A(\cdot\wedge\zeta_R)$, respectively:

$$(X(t\wedge\zeta_R)^2-\int_0^{t\wedge\zeta_R}X(s)X(ds),\mathcal{F}_t,P)$$

is a martingale; and therefore

$$E^P[X(t\wedge\zeta_R)^2] = E^P\left[\int_0^{t\wedge\zeta_R}X(s)X(ds)\right].$$

On the other hand, since $X(\cdot\wedge\zeta_R)$ is P-almost surely continuous and of local bounded variation, $X(t\wedge\zeta_R)^2 = 2\int_0^{t\wedge\zeta_R}X(s)X(ds)$ (a.s.,P), and therefore $E^P[X(t\wedge\zeta_R)^2] = 2E^P\left[\int_0^{t\wedge\zeta_R}X(s)X(ds)\right]$. Hence, $E^P[X(t\wedge\zeta_R)^2] = 0$ for all $t \geq 0$, and so $X(t\wedge\zeta_R) = 0$, $t \geq 0$, (a.s.,P). Clearly this leads immediately to the conclusion that $X(t\wedge\zeta_R) = 0$, $t \geq 0$, (a.s.,P).

To prove the last assertion, it suffices to check that, for each $0 \leq s < t$, $|X_s|(t) = \infty$ (a.s.,P), where $X_s(\cdot) \equiv X(\cdot) - X(\cdot\wedge s)$. But, if $\zeta^s \equiv \sup\{u\geq 0: |X_s|(u) < \infty\}$, then $P(|X_s|(t)<\infty) = P(\zeta^s>t)$; and, by the preceding with X_s replacing X, $P(\zeta^s>t) \leq P(X(t)=X(s)) = 0$. Q.E.D.

(2.24) <u>Corollary</u>: Let $X: [0,\infty)\times E \longrightarrow \mathbb{R}^1$ be a right continuous progressively measurable function. Then there is, up to a P-null set, at most one right continuous progressively measurable $A: [0,\infty)\times E \longrightarrow \mathbb{R}^1$ such that: $A(0) \equiv 0$, $t \longmapsto A(t)$ is P-almost surely continuous and of local bounded variation, and, in addition, $(X(t)-A(t),\mathcal{F}_t,P)$ is a martingale.

<u>Proof</u>: Suppose that there were two, A and A'. Then $(A(t)-A'(t),\mathcal{F}_t,P)$ would be a P-almost surely continuous martingale which is P-almost surely of local bounded variation. Hence, by Theorem (2.23), we would have $A(t) - A'(t) = A(0) - A'(0) = 0$, $t \geq 0$, (a.s.,P). Q.E.D.

Before proving the existence part of our special case of the Doob-Meyer decomposition theorem, we mention a result which addresses an extremely pedantic issue. Namely, for technical reasons (e.g. countability considerations), it is often better not to complete the σ-algebras $\mathcal{F}_t$ with respect to P. At the same time, it is convenient to have the processes under consideration right continuous for every $\xi \in E$, not just P-almost every one. In order to make sure that we can make our processes everywhere right continuous and, at the same time, progressively measurable with respect to possibly incomplete σ-algebras $\mathcal{F}_t$, we will sometimes make reference to the following lemma, whose proof may be found in 4.3.3 of [S.&V.]. On the other hand, since in most cases there is either no harm in completing the σ-algebras or the asserted conclusion is clear from other considerations, we will not bother with the proof here.

(2.25) <u>Lemma</u>: Let $\{X_n : n\geq 1\}$ be a sequence of right continuous (P-almost surely continuous) progressively measurable functions with values in a Banach space $(B,\|\cdot\|)$. If

$$\lim_{m\to\infty} \sup_{n\geq m} P(\sup_{0\leq t\leq T}\|X_n(t)-X_m(t)\|\geq\epsilon) = 0$$

for every $T > 0$ and $\epsilon > 0$, then there is a P-almost surely unique right-continuous (P-almost surely continuous) progressively measurable function X such that

$$\lim_{n\to\infty} P(\sup_{0\leq t\leq T}\|X_n(t)-X(t)\|\geq\epsilon) = 0$$

for all $T > 0$ and $\epsilon > 0$.

(2.26) <u>Theorem</u> (Doob-Meyer): Let $(X(t),\mathcal{F}_t,P)$ be a P-almost surely continuous real-valued $L^2(P)$-martingale. Then there is a P-almost surely unique right continuous progressively measurable $A: [0,\infty)\times E\longrightarrow[0,\infty)$ such that: $A(0) \equiv 0$, $t\longmapsto A(t)$ is non-decreasing and P-almost surely continuous and $(X(t)^2-A(t),\mathcal{F}_t,P)$ is a martingale.

<u>Proof</u>: The uniqueness is clearly a consequence of Corollary (2.24). In proving existence, we assume, without loss in generality, that $X(0) \equiv 0$.

Define $\tau_k^0 \equiv k$ for $k \geq 0$; and given $n \geq 1$, define $\tau_0^n \equiv 0$ and for $\ell \geq 0$:

$$\tau_{\ell+1}^n = (\inf\{t\geq\tau_\ell^n: \sup_{\tau_\ell^n\leq s\leq t}|X(s)-X(\tau_\ell^n)|\geq 1/n\})\wedge(\tau_\ell^n+1/n)\wedge\tau_{k+1}^{n-1},$$

if $\tau_k^{n-1} \leq \tau_\ell^n < \tau_{k+1}^{n-1}$. Clearly the τ_k^n's are stopping times, $\tau_k^n < \tau_{k+1}^n$, and $\{\tau_k^n\} \subseteq \{\tau_k^{n+1}\}$ (a.s.,P). In addition, by P-almost sure continuity, $\tau_k^n \longrightarrow \infty$ (a.s.,P) as $k \longrightarrow \infty$ and $|X(\tau_{k+1}^n) - X(\tau_k^n)| \le 1/n$, $k \ge 0$, (a.s.,P). Choose $K_0 < \cdots < K_n < \cdots$ so that $P(\Lambda_n^c) \le 1/n$ where $\Lambda_n = \{\tau_{K_n}^n > n\}$, and define:

$$M_n(t) = \sum_{k=0}^{K_n} X(\tau_k^n)(X(t\wedge\tau_{k+1}^n) - X(t\wedge\tau_k^n))$$

and

$$A_n(t) = \sum_{k=0}^{K_n} (X(t\wedge\tau_{k+1}^n) - X(t\wedge\tau_k^n))^2.$$

Clearly, for all $n \ge 0$: $M_n(0) = A_n(0) \equiv 0$; $(M_n(t),\mathcal{F}_t,P)$ is a P-almost surely continuous martingale; A_n is a right-continuous, non-negative, progressively measurable function which is P-almost surely continuous; and $A_n(s) \le A_n(t)$ if $t \ge s + 1/n$. Moreover, for $n \ge 1$, $0 \le t \le n$, and $\xi \in \Lambda_n$:

$$X(t,\xi)^2 = 2M_n(t,\xi) + A_n(t,\xi). \tag{2.27}$$

Given $T > 0$ and $\epsilon > 0$, we are now going to show that

$$\lim_{m\longrightarrow\infty} \sup_{n\ge m} P(\sup_{0\le t\le T}|M_n(t)-M_m(t)|\ge\epsilon) = 0.$$

To this end, let $T \le m < n$ and set $\zeta = T\wedge\tau_{K_m}^m\wedge\tau_{K_n}^n$. Then:

$$\begin{aligned} &P(\sup_{0\le t\le T}|M_n(t)-M_m(t)|\ge\epsilon) \\ &\quad\le P(\Lambda_m^c\cup\Lambda_n^c) + P(\sup_{0\le t\le T}|M_n(t\wedge\zeta)-M_m(t\wedge\zeta)|\ge\epsilon) \\ &\quad\le 2/m + 1/\epsilon^2 E^P[(M_n(\zeta)-M(\zeta))^2], \end{aligned}$$

where we have used Doob's inequatity. Define $\rho_k = \tau_k^m\wedge\zeta$ and $\sigma_\ell = \tau_\ell^n\wedge\zeta$. Note that:

$$M_n(\zeta)-M_m(\zeta)$$

$$= \sum_{k=0}^{\infty} \sum_{\ell=0}^{\infty} \chi_{[\rho_k,\rho_{k+1})}(\sigma_\ell)(X(\sigma_\ell)-X(\rho_k))(X(\sigma_{\ell+1})-X(\sigma_\ell))$$

(a.s.,P) and that the terms in this double sum are mutually P-orthognal. Hence:

$$E^P[(M_n(\zeta)-M(\zeta))^2]$$

$$= E^P\Big[\sum_{k=0}^{\infty} \sum_{\ell=0}^{\infty} \chi_{[\rho_k,\rho_{k+1})}(\sigma_\ell)(X(\sigma_\ell)-X(\rho_k))^2(X(\sigma_{\ell+1})-X(\sigma_\ell))^2\Big]$$

$$\leq 1/m^2E^P\Big[\sum_{k=0}^{\infty} \sum_{\ell=0}^{\infty} \chi_{[\rho_k,\rho_{k+1})}(\sigma_\ell)(X(\sigma_{\ell+1})-X(\sigma_\ell))^2\Big]$$

$$= 1/m^2E^P\Big[\sum_{\ell=0}^{\infty}(X(\sigma_{\ell+1})-X(\sigma_\ell))^2\Big] = 1/m^2E^P[X(\zeta)^2]$$

$$\leq 1/m^2E^P[X(T)^2],$$

where we have used the P-orthognality of $\{X(\sigma_{\ell+1})-X(\sigma_\ell)\}$ in the last line.

We now apply Lemma (2.25) and conclude that there is a right continuous, P-almost surely continuous, progressively measurable $M: [0,\infty)\times E\longrightarrow\mathbb{R}^1$ such that $M_n\longrightarrow M$, uniformly on finite intervals, in P-measure. Furthermore, our argument shows that $M_m(t)\longrightarrow M(t)$ in $L^1(P)$ for each $t \geq 0$. Hence, $(M(t),\mathcal{F}_t,P)$ is a P-almost surely continuous martingale.

Finally, set $A' = X^2 - 2M$. Then, as a consequence of (2.27), we see that $A_n\longrightarrow A'$, uniformly on finite intervals, in P-measure. In particular, $A'(0) = 0$ and $t\longmapsto A'(t)$ is non-decreasing P-almost surely. Thus, we are done once we define $A(t) = \sup_{0<s<t}(A'(s)-A'(0))$ for $t \geq 0$. Q.E.D.

(2.28) Exercise: Let $(\beta(t), \mathcal{F}_t, P)$ be a one-dimensional Wiener process. Show that $(\beta(t), \mathcal{F}_t, P)$ is an $L^2(P)$-martingale and that the associated non-decreasing process constructed in the preceding is $A(t) \equiv t$, $t \geq 0$. In addition, show that for each $T \geq 0$, $\sum_{k=0}^{2^n-1} (\beta((k+1)T/2^n) - \beta(kT/2^n))^2 \longrightarrow T$ (a.s.,P) as $n \longrightarrow \infty$.

Let $Mart_c^2$ (= $Mart_c^2(\{\mathcal{F}_t\}, P)$) denote the space of all real-valued P-almost surely continuous $L^2(P)$-martingales $(X(t), \mathcal{F}_t, P)$. Clearly $Mart_c^2$ is a linear space. Given $X \in Mart_c^2$, we will use $\langle X \rangle$ to denote the associated process A constructed in Theorem (2.26). In addition, given $X, Y \in Mart_c^2$, define $\langle X, Y \rangle$ by:

$$\langle X,Y \rangle = 1/4[\langle X+Y \rangle - \langle X-Y \rangle].$$

Clearly, $\langle X,Y \rangle$ is a right-continuous progressively measurable function which is not only of local bounded variation, with $|\langle X,Y \rangle|(T) \in L^1(P)$ for all $T \geq 0$, but also P-almost surely continuous.

(2.29) Exercise: Given a stopping time τ and an $X \in Mart_c^2$, define $X^\tau(t) = X(t \wedge \tau)$ and $X_\tau(t) = X(t) - X^\tau(t)$, $t \geq 0$. Show that X^τ and X_τ are elements of $Mart_c^2$ and that $\langle X^\tau \rangle(t) = \langle X \rangle(t \wedge \tau)$ and $\langle X_\tau \rangle(t) = \langle X \rangle(t) - \langle X \rangle(t \wedge \tau)$, $t \geq 0$, (a.s.,P).

(2.30) Theorem(Kunita & Watanabe): Given $X, Y \in Mart_c^2$, $\langle X,Y \rangle$ is the P-almost surely unique right continuous progressively measurable function which has local bounded variation, is P-almost surely continuous, and has the properties that $\langle X,Y\rangle(0) \equiv 0$ and $(X(t)Y(t)-\langle X,Y\rangle(t),\mathcal{F}_t,P)$ is a martingale. In particular, $\langle X\rangle = \langle X,X\rangle$ (a.s.,P) for all X $\in \text{Mart}_c^2$; and, for all $X,Y,Z \in \text{Mart}_c^2$, $\langle aX+bY,Z\rangle = a\langle X,Z\rangle + b\langle Y,Z\rangle$, $a,b \in \mathbb{R}^1$, (a.s.,P). Finally, for all $X,Y \in \text{Mart}_c^2$: $|\langle X,Y\rangle|(\Gamma) \le \langle X\rangle(\Gamma)^{1/2}\langle Y\rangle(\Gamma)^{1/2}$, $\Gamma \in \mathcal{B}_{[0,\infty)}$, (a.s.,P); $\langle X+Y\rangle(\Gamma) \le \langle X\rangle(\Gamma)^{1/2} + \langle Y\rangle(\Gamma)^{1/2}$, $\Gamma \in \mathcal{B}_{[0,\infty)}$, (a.s.,P); and $|\langle X,Y\rangle|(dt) \le (\langle X\rangle(dt) + \langle Y\rangle(dt))/2$ (a.s.,P).

Proof: To prove the first assertion, simply note that $XY = 1/4[(X+Y)^2 - (X-Y)^2]$, and apply Lemma (2.24).

The equality $\langle X\rangle = \langle X,X\rangle$ as well as the linearity assertion follow easily from uniqueness.

In order to prove the rest of the theorem, it suffices to show that $|\langle X,Y\rangle|(\Gamma) \le \langle X\rangle(\Gamma)^{1/2}\langle Y\rangle(\Gamma)^{1/2}$, $\Gamma \in \mathcal{B}_{[0,\infty)}$, (a.s.,P); and to do this we need only check that, for each $0 \le s < t$,

$$|\langle X,Y\rangle(t) - \langle X,Y\rangle(s)|$$
$$\le (\langle X\rangle(t) - \langle X\rangle(s))^{1/2}(\langle Y\rangle(t) - \langle Y\rangle(s))^{1/2} \text{ (a.s.,P)}.$$

Furthermore, by replacing X and Y with X_s and Y_s, respectively (cf. Exercise (2.29)), we see that it is enough to prove that $|\langle X,Y\rangle(t)| \le \langle X\rangle(t)^{1/2}\langle Y\rangle(t)^{1/2}$ (a.s.,P) for each $t \ge 0$. But, by the linearity property, $0 \le \langle \lambda X \pm Y/\lambda\rangle(t) = \lambda^2\langle X\rangle(t) \pm 2\langle X,Y\rangle(t) + \langle Y\rangle(t)/\lambda^2$, $\lambda > 0$, (a.s.,P). Hence the desired inequality follows by the same argument as one uses to prove the ordinary Schwarz's inequality. Q.E.D.

(2.31) <u>Exercise</u>: Given $X, Y \in \text{Mart}_c^2(\{\mathcal{F}_t\}, P)$ and an $\{\mathcal{F}_t\}$-stopping time τ, show that

$$\langle X^\tau, Y\rangle(\cdot) = \chi_{[0,\tau)}(\cdot)\langle X, Y\rangle(\cdot).$$

Next, set $\mathcal{F}_t^X = \sigma(X(s): 0 \le s \le t)$ and $\mathcal{F}_t^Y = \sigma(Y(s): 0 \le s \le t)$. Show that $X, Y \in \text{Mart}_c^2(\{\mathcal{F}_t^X \times \mathcal{F}_t^Y\}, P)$ and that, up to a P-null set, $\langle X, Y\rangle$ defined relative to $(\{\mathcal{F}_t^X \times \mathcal{F}_t^Y\}, P)$ coincides with $\langle X, Y\rangle$ defined relative to $(\{\mathcal{F}_t\}, P)$. Conclude that, if for some $T > 0$, $\mathcal{F}_T^X$ and $\mathcal{F}_T^Y$ are P-independent, then $\langle X, Y\rangle(t) = 0$, $0 \le t \le T$, (a.s., P).

3. Stochastic Integrals, Itô's Formula, and Semi-martingales:

We continue with the notation with which we were working in section 2.

Given a right continuous, non-decreasing, P-almost surely continuous, progressively measurable function A: $[0,\infty)\times E\longrightarrow[0,\infty)$, denote by $\underline{L^2_{loc}(A,P)} = \underline{L^2_{loc}(\{\mathcal{F}_t\},A,P)}$ the space of progressively measurable α: $[0,\infty)\times E\longrightarrow\mathbb{R}^1$ such that $E^P\left[\int_0^T\alpha(t)^2A(dt)\right] < \infty$ for all $T > 0$. Clearly, $L^2_{loc}(A,P)$ admits a natural metric with respect to which it becomes a Frechet space.

Given $X \in Mart^2_c$ and $\alpha \in L^2_{loc}(\langle X\rangle,P)$, note that there is at most one I: $[0,\infty)\times E\longrightarrow\mathbb{R}^1$ such that:

$$\begin{aligned} &\text{i) } I(0) \equiv 0 \text{ and } I \in Mart^2_c, \\ &\text{ii) } \langle I,Y\rangle = \alpha\langle X,Y\rangle \text{ (a.s.,P) for all } Y \in Mart^2_c. \end{aligned} \tag{3.1}$$

(Given a measure μ and a measurable function α, $\underline{\alpha\mu}$ denotes the measure ν such that $\frac{d\nu}{d\mu} = \alpha$.) Indeed, if there were two, say I and I', then we would have $\langle I-I',Y\rangle \equiv 0$ (a.s.,P) for all $Y \in Mart^2_c$. In particular, taking $Y = I - I'$, we would conclude that $E^P[(I(T)-I'(T))^2] = 0$, $T \geq 0$, and therefore that $I = I'$ (a.s.,P). Following Kunita and Watanabe, we will say that, if it exists, the P-almost surely unique I satisfying (3.1) is the (Itô) stochastic integral of α with respect to X and we will denote I by $\underline{\int_0^\cdot\alpha(s)dX(s)}$.

Observe that if $\alpha,\beta \in L^2_{loc}(\langle X\rangle,P)$ and if both $\int_0^\cdot\alpha(s)dX(s)$ and $\int_0^\cdot\beta(s)dX(s)$ exist, then: $\int_0^\cdot[a\alpha(s)+b\beta(t)]dX(s)$

exists and is equal $a\int_0^{\cdot}\alpha(s)dX(s) + b\int_0^{\cdot}\beta(s)dX(s)$ (a.s.,P) for all $a,b \in \mathbb{R}^1$, and

$$E^P\Big[\sup_{0\le t\le T}\Big(\int_0^t \alpha(s)dX(s)-\int_0^t \beta(s)dX(s)\Big)^2\Big]$$
$$\le 4E^P\Big[\Big(\int_0^T \alpha(s)dX(s)-\int_0^T \beta(s)dX(s)\Big)^2\Big] \qquad (3.2)$$
$$= 4E^P\Big[\int_0^T (\alpha(t)-\beta(t))^2\langle X\rangle(dt)\Big],\ T \ge 0.$$

From this it is easy to see that the set of α's for which $\int_0^{\cdot}\alpha(s)dX(s)$ exists is a closed linear subspace of $L^2_{loc}(\langle X\rangle,P)$.

(3.3) Exercise: Let $X \in Mart^2_c$ be given and suppose that $\sigma \le \tau$ are stopping times. Let γ be an $\mathcal{F}_\sigma$-measurable function satisfying $E^P[\gamma^2(\langle X\rangle(T\wedge\tau)-\langle X\rangle(T\wedge\sigma))] < \infty$ for all $T > 0$, and set $\alpha(t) = \chi_{[\sigma,\tau)}(t)\gamma$, $t \ge 0$. Show that $\alpha \in L^2_{loc}(\langle X\rangle,P)$ and that $\int_0^{\cdot}\alpha(s)dX(s)$ exists and is equal to $\gamma(X^\tau(\cdot)-X^\sigma(\cdot))$ (see exercise (2.29) for the notation here).

We want to show that $\int_0^{\cdot}\alpha(s)dX(s)$ exists for all $\alpha \in L^2_{loc}(\langle X\rangle,P)$. To this end, first suppose that α is simple in the sense that there is an $n \ge 1$ for which $\alpha(t) = \alpha([nt]/n)$, $t \ge 0$. Set:

$$I(t) = \sum_{k=0}^{\infty} \alpha(k/n)(X(t\wedge((k+1)/n))-X(t\wedge(k/n))),\ t \ge 0.$$

Then $I \in Mart^2_c$. Moreover, if $k/n \le s \le t \le (k+1)/n$, then:

$$\begin{aligned} E^P[I(t)Y(t)-I(s)Y(s)|\mathcal{F}_s] &= E^P[(I(t)-I(s))(Y(t)-Y(s))|\mathcal{F}_s] \\ &= \alpha(k/n)E^P[(X(t)-X(s))(Y(t)-Y(s))|\mathcal{F}_s] \\ &= \alpha(k/n)E^P[\langle X,Y\rangle(t) - \langle X,Y\rangle(s)|\mathcal{F}_s] \\ &= E^P\left[\int_s^t \alpha(u)\langle X,Y\rangle(du)|\mathcal{F}_s\right] \text{ (a.s.,P)}. \end{aligned}$$

In other words, $\langle I,Y\rangle = \alpha\langle X,Y\rangle$ (a.s.,P), and so $\int_0^{\cdot}\alpha(s)dX(s)$ exists and is given by I. Knowing that $\int_0^{\cdot}\alpha(s)dX(s)$ exists for simple α's, one can easily show that $\int_0^{\cdot}\alpha(s)dX(s)$ exists for bounded progressively measurable α's which are P-almost surely continuous; indeed, simply take $\alpha_n(t) = \alpha([nt]/n)$, $t \geq 0$, and note that $\alpha_n \longrightarrow \alpha$ in $L^2_{loc}(\langle X\rangle,P)$. Thus, we will have completed our demonstration that $\int_0^{\cdot}\alpha(s)dX(s)$ exists for all $\alpha \in L^2_{loc}(\langle X\rangle,P)$ once we have proved the following approximation result.

(3.4) <u>Lemma</u>: Let $A: [0,\infty)\times E \longrightarrow [0,\infty)$ be a non-decreasing, P-almost surely continuous, progressively measurable function with $A(0) \equiv 0$. Given $\alpha \in L^2_{loc}(A,P)$, there is a sequence $\{\alpha_n\} \subseteq L^2_{loc}(A,P)$ of bounded, P-almost surely continuous functions which tend to α in $L^2_{loc}(A,P)$.

<u>Proof</u>: Since the space of bounded elements of $L^2_{loc}(A,P)$ are obviously dense in $L^2_{loc}(A,P)$, we will assume that α is bounded.

We first handle the special case when $A(t) \equiv t$ for all $t \geq 0$. To this end, choose $\rho \in C_0^\infty((0,1))$ so that $\int\rho(t)dt = 1$, and extend α to $\mathbb{R}^1\times E$ by setting $\alpha(t) \equiv 0$ for $t < 0$. Next, define $\alpha_n(t) = n\int \alpha(t-s)\rho(ns)ds$ for $t \geq 0$ and $n \geq 1$. Then it is easy to check that $\{\alpha_n\}$ will serve.

To handle the general case, first note that it suffices for us to show that for each $T > 0$ and $\epsilon > 0$ there exists a bounded P-almost surely continuous $\alpha' \in L^2_{loc}(A,P)$ such that $E^P\left[\int_0^T (\alpha(t)-\alpha'(t))^2 A(dt)\right] < \epsilon^2$. Given T and ϵ, choose $M > 1$ so that $E^P\left[\int_0^T \alpha(t)^2 A(dt),\ A(T) \geq M-1\right] < (\epsilon/2)^2$ and $\eta \in C^\infty(\mathbb{R}^1)$ so that $\chi_{[0,M-1]} \leq \eta \leq \chi_{[-1,M]}$. Set $B(t) = \int_0^t \eta(A(s))^2 A(ds) + t$, $t \geq 0$, and $\tau(t) = B^{-1}(t)$. Then $\{\tau(t): t \geq 0\}$ is a non-decreasing family of bounded stopping times. Set $\mathcal{G}_t = \mathcal{F}_{\tau(t)}$ and $\beta(t) = \alpha(\tau(t))$. Then β is a bounded $\{\mathcal{G}_t\}$-progressively measurable function; and so, by the preceding, we can find a bounded continuous $\{\mathcal{G}_t\}$-progressively measurable β' such that

$$E^P\left[\int_0^{T+M} (\beta(t)-\beta'(t))^2 dt\right] < (\epsilon/2)^2.$$

Finally, define $\alpha'(t) = \beta'(B(t))\eta(A(t))$. Then α' is a bounded P-almost surely continuous element of $L^2_{loc}(A,P)$, and:

$$\begin{aligned} E^P\left[\int_0^T (\alpha(t)-\alpha'(t))^2 A(dt)\right]^{1/2} &\leq \epsilon/2 + E^P\left[\int_0^T (\alpha(t)-\beta'(B(t)))^2 B(dt)\right]^{1/2} \\ &\leq \epsilon/2 + E^P\left[\int_0^{T+M} (\beta(t)-\beta'(t))^2 dt\right]^{1/2} < \epsilon. \end{aligned}$$

Q.E.D.

As a consequence of the preceding, we now know that

$\int_0^\cdot \alpha(s)dX(s)$ exists for all $X \in \text{Mart}_c^2$ and $\alpha \in L^2_{\text{loc}}(\langle X\rangle,P)$.

(3.5) <u>Exercise</u>: Let $X \in \text{Mart}_c^2$ be given.

i) Given stopping times $\sigma \le \tau$ and $\alpha \in L^2_{\text{loc}}(\langle X\rangle,P)$, show that $\chi_{[\sigma,\tau)}(t)\alpha(t) \in L^2_{\text{loc}}(\langle X\rangle,P)$ and that:

$$\int_{T\wedge\sigma}^{T\wedge\tau}\alpha(t)dX(t) \equiv \int_0^T \chi_{[\sigma,\tau)}\alpha(t)dX(t)$$

$$= \int_0^{T\wedge\tau}\alpha(t)dX(t) - \int_0^{T\wedge\sigma}\alpha(t)dX(t) \quad (\text{a.s.},P)$$

ii) Given $\beta \in L^2_{\text{loc}}(\langle X\rangle,P)$ and $\alpha \in L^2_{\text{loc}}(\beta^2\langle X\rangle,P)$, show that:

$$\int_0^T \alpha(t)d\left(\int_0^t \beta(s)dX(s)\right) = \int_0^T \alpha(s)\beta(s)dX(s) \quad (\text{a.s.},P).$$

Our next project is the derivation of the reknowned <u>Itô's formula</u>. (Our presentation again follows that of Kunita and Watanabe). Namely, let $X = (X^1,\dots,X^M) \in (\text{Mart}_c^2)^M$ and $Y\colon [0,\infty)\times E \longrightarrow \mathbb{R}^N$ be a P-almost surely continuous progressively measurable function of local bounded variation such that $|Y|(T) \equiv \left(\sum_1^N |Y^j|(T)^2\right)^{1/2} \in L^1(P)$ for each $T > 0$. Given $f \in C_b^{2,1}(\mathbb{R}^M\times\mathbb{R}^N)$, Itô's formula is the statement that:

$$f(Z(T)) - f(Z(0))$$

$$= \sum_{i=1}^M \int_0^T \partial_{x^i} f(Z(t))dX^i(t) + \sum_{j=1}^N \int_0^T \partial_{y^j} f(Z(t))Y^j(dt) \tag{3.6}$$

$$+ 1/2 \sum_{i,i'=1}^M \int_0^T \partial_{x^i}\partial_{x^{i'}} f(Z(t))\langle X^i,X^{i'}\rangle(dt) \quad (\text{a.s.},P),$$

where $Z \equiv (X,Y)$.

It is clear that, since (3.6) is just an identification statement, we may assume that $t \longmapsto Z(t,\xi)$ and $t \longmapsto \underline{\langle\langle X,X\rangle\rangle}(t,\xi) \equiv ((\langle X^i,X^{i'}\rangle(t,\xi)))_{i,i'=1}^{M}$ are continuous for all $\xi \in E$. In addition, it suffices to prove (3.6) when $f \in C_b^\infty(\mathbb{R}^{M+N})$. Thus, we will proceed under these assumptions. Given $n \geq 1$, define τ_k^n, $k \geq 0$, so that $\tau_0^n \equiv 0$ and

$$\tau_{k+1}^n = [\inf\{t \geq \tau_k^n : (\max_{1\leq i\leq M}(\langle X^i\rangle(t) - \langle X^i\rangle(\tau_k^n)) \vee |Z(t) - Z(\tau_k^n)| \geq 1/n\}] \wedge (\tau_k^n + 1/n) \wedge T.$$

Then, for each $T > 0$ and $\xi \in E$, $\tau_k^n(\xi) = T$ for all but a finite number of k's. Hence, $f(Z(T)) - f(Z(0)) = \sum_{k=0}^\infty (f(Z_{k+1}^n) - f(Z_k^n))$, where $Z_k^n = (X_k^n, Y_k^n) \equiv Z(\tau_k^n)$. Clearly:

$$f(Z_{k+1}^n) - f(Z_k^n) =$$

$$[f(X_{k+1}^n, Y_k^n) - f(X_k^n, Y_k^n)] + [f(X_{k+1}^n, Y_{k+1}^n) - f(X_{k+1}^n, Y_k^n)]$$

$$= \sum_{i=1}^M \partial_{x^i} f(Z_k^n)\Delta_k^n X_i + 1/2 \sum_{i,i'=1}^M \int_0^T \partial_{x^i}\partial_{x^{i'}} f(Z_k^n))\Delta_k^n \langle X^i, X^{i'}\rangle$$

$$+ \sum_{j=1}^N \int_{\tau_k^n}^{\tau_{k+1}^n} \partial_{y_j} f(X_{k+1}^n, Y(t)) Y_j(dt) + R_k^n,$$

where $\Delta_k^n \Xi \equiv \Xi(\tau_{k+1}^n) - \Xi(\tau_k^n)$ and

$$R_k^n \equiv 1/2 \sum_{i,i'=1}^M (\partial_{x^i}\partial_{x^{i'}} f(\hat{Z}_k^n) - \partial_{x^i}\partial_{x^{i'}} f(Z_k^n))\Delta_k^n X^i \Delta_k^n X^{i'}$$

$$+ 1/2 \sum_{i,i'=1}^M \partial_{x^i}\partial_{x^{i'}} f(Z_k^n)(\Delta_k^n X^i \Delta_k^n X^{i'} - \Delta_k^n \langle X^i, X^{i'}\rangle)$$

with $\hat{Z}_k^n$ a point on the line joining (X_{k+1}^n, Y_k^n) to (X_k^n, Y_k^n). By exercise (3.3),

$$\sum_{k=0}^{\infty} \partial_{x^i} f(Z_k^n)\Delta_k^n X^i = \int_0^T \partial_{x^i} f(Z^n(s))dX^i(s)$$

where $Z^n(s) \equiv Z_k^n$ for $s \in [\tau_k^n, \tau_{k+1}^n)$ and $Z^n(s) \equiv Z(T)$ for $s \geq T$. Since $Z^n(s) \longrightarrow Z(s)$ uniformly for $s \in [0,T]$, we conclude that

$$\sum_{k=0}^{\infty} \partial_{x^i} f(Z_k^n)\Delta_k^n X^i \longrightarrow \int_0^T \partial_{x^i} f(Z(s))dX^i(s)$$

in $L^2(P)$. Also, from standard integration theory

$$\sum_{k=0}^{\infty}\int_0^T \partial_{x^i}\partial_{x^{i'}} f(Z_k^n))\Delta_k^n\langle X^i,X^{i'}\rangle \longrightarrow$$

$$\sum_{k=0}^{\infty}\int_0^T \partial_{x^i}\partial_{x^{i'}} f(Z(s)))\langle X^i,X^{i'}\rangle(ds)$$

and

$$\sum_{k=0}^{\infty}\int_{\tau_k^n}^{\tau_{k+1}^n} \partial_{y^j} f(X_{k+1}^n, Y(t))Y^j(dt) \longrightarrow \int_0^T \partial_{y^j} f(Z(s))Y^j(ds)$$

in $L^1(P)$. It therefore remains only to check that $\sum_k R_k^n \longrightarrow 0$ in P-measure.

First observe that

$$|(\partial_{x^i}\partial_{x^{i'}} f(\hat{Z}_k^n) - \partial_{x^i}\partial_{x^{i'}} f(Z_k^n))\Delta_k^n X^i \Delta_k^n X^{i'}|$$
$$\leq C[(\Delta_k^n X^i)^2 + (\Delta_k^n X^{i'})^2]/n$$

and therefore that

$$E^P[|\sum_k(\partial_{x^i}\partial_{x^{i'}} f(\hat{Z}_k^n) - \partial_{x^i}\partial_{x^{i'}} f(Z_k^n))\Delta_k^n X^i \Delta_k^n X^{i'})|]$$
$$\leq 2CE^P[|X(T) - X(0)|^2]/n \longrightarrow 0.$$

At the same time:

$$E^P[(\sum_{k=0}^{\infty} \partial_{x^i}\partial_{x^{i'}} f(Z_k^n)(\Delta_k^n X^i \Delta_k^n X^{i'} - \Delta_k^n\langle X^i,X^{i'}\rangle))^2]$$
$$= \sum_{k=0}^{\infty} E^P[(\partial_{x^i}\partial_{x^{i'}} f(Z_k^n)(\Delta_k^n X^i \Delta_k^n X^{i'} - \Delta_k^n\langle X^i,X^{i'}\rangle))^2]$$
$$\leq C\sum_{k=0}^{\infty} E^P[((\Delta_k^n X^i \Delta_k^n X^{i'} - \Delta_k^n\langle X^i,X^{i'}\rangle))^2]$$

$$\leq C' \sum_{k=0}^{\infty} E^P[(\Delta_k^n X^i)^4 + (\Delta_k^n X^{i'})^4 + (\Delta_k^n \langle X^i \rangle)^2 + (\Delta_k^n \langle X^{i'} \rangle)^2]$$

$$\leq C'' \sum_{k=0}^{\infty} E^P[|\Delta_k^n X|^2]/n = C''E^P[|X(T)-X(0)|^2]/n \longrightarrow 0.$$

Combining these, we now see that (3.6) holds.

The applications of Itô's formula are innumerable. One particularly beautiful one is the following derivation, due to Kunita and Watanabe, of a theorem proved originally by Levy.

(3.7) Theorem (Levy): Let $\beta \in (\mathrm{Mart}_c^2)^N$ and assume that $\langle\langle \beta, \beta \rangle\rangle(t) = tI$, $t \geq 0$ (i.e. $\langle \beta^i, \beta^j \rangle(t) = t\delta^{i,j}$). Then $(\beta(t)-\beta(0), \sigma(\beta(s): 0 \leq s \leq t), P)$ is an N-dimensional Wiener process.

Proof: We assume, without loss of generality, that $\beta(0) \equiv 0$. What we must show is that $P \circ \beta^{-1} = \mathscr{W}$; and, by Corollary (1.13), this comes down to showing that

$$(\varphi(x(t)) - 1/2\int_0^t \Delta\varphi(x(s))ds, \mathscr{M}_t, P\circ\beta^{-1})$$

is a martingale for every $\varphi \in C_0^\infty(\mathbb{R}^N)$. Clearly this will follow if we show that $(\varphi(\beta(t)) - 1/2\int_0^t \Delta\varphi(\beta(s))ds, \mathscr{F}_t, P)$ is a martingale. But, by Itô's formula:

$$\varphi(\beta(t)) - \varphi(\beta(0)) - 1/2\int_0^t \Delta\varphi(\beta(s))ds = \sum_{i=1}^N \int_0^t \partial_{x^i}\varphi(\beta(s))d\beta^i(s),$$

and so the proof is complete. Q.E.D.

Given a right continuous, P-almost surely continuous,

$\{\mathcal{F}_t\}$-progressively measurable function $\beta\colon [0,\infty)\times E\longrightarrow\mathbb{R}^N$, we will say that $\underline{(\beta(t),\mathcal{F}_t,P)}$ is an $\underline{N}$ -dimensional Brownian motion if $\beta \in (\mathrm{Mart}_c^2(\{\mathcal{F}_t\},P))^N$, $\beta(0) = 0$, and $\langle\langle\beta,\beta\rangle\rangle(t) \equiv t$, $t \geq 0$, (a.s.,P).

(3.8) Exercise:

i) Let $\beta\colon [0,\infty)\times E\longrightarrow\mathbb{R}^N$ be a right continuous, P-almost surely continuous, progressively measurable function with $\beta(0) \equiv 0$. Show that $(\beta(t),\mathcal{F}_t,P)$ is an N-dimensional Brownian motion if and only if $P(\beta(t)\in\Gamma|\mathcal{F}_s) = \int_\Gamma g(t-s,Y-\beta(s))dy$, $0 \leq s \leq t$ and $\Gamma \in \mathcal{B}_{R^N}$, where $g(\cdot,*)$ denotes the N-dimensional Gauss kernel.

ii) Generalize Levy's theorem by showing that if a and b are as in section 1 and $\{P_{s,x}\colon (s,x)\in[0,\infty)\times\mathbb{R}^N\}$ is the associated family of measures on Ω, then, for each (s,x), $P_{s,x}$ is the unique $P \in \mathbf{M}_1(\Omega)$ such that:

$$P(x(0)=x) = 1,$$

$$x(\cdot)-\int_0^{\cdot} b(s+t,x(t))dt \in (\mathrm{Mart}_c^2(\{\mathcal{M}_t\},P))^N,$$

and

$$\langle\langle x(\cdot)-\int_0^{\cdot} b(s+t,x(t))dt,x(\cdot)-\int_0^{\cdot} b(s+t,x(t))dt\rangle\rangle(T)$$
$$= \int_0^T a(s+t,x(t))dt,\ T \geq 0,\ (\text{a.s.},P)$$

Although the class Mart_c^2 has many pleasing properties, it is not invariant under changes of coordinates. (That is, even if $f \in C^\infty(\mathbb{R}^1)$, $f\circ X$ will seldom be an element of Mart_c^2

simply because X is.) There are two reasons for this, the first of which is the question of integrability. To remove this first problem, we introduce the class $\underline{\text{Mart}}_c^{loc}$ (= $\underline{\text{Mart}}_c^{loc}(\{\underline{\mathcal{F}}_t, \underline{P}\})$ of P-almost surely continuous local martingales. Namely, we say that $X \in \text{Mart}_c^{loc}$ if $X: [0,\infty)\times E \longrightarrow \mathbb{R}^1$ is a right continuous, P-almost surely continous function for which there exists a non-decreasing sequence of stopping times σ_n with the properties that $\sigma_n \longrightarrow \infty$ (a.s.,P) and $(X^{\sigma_n}(t), \mathcal{F}_t, P)$ is a bounded martingale for each n (recall that $X^{\sigma}(\cdot) \equiv X(\cdot\wedge\sigma))$. It is easy to check that Mart_c^{loc} is a linear space. Moreover, given $X \in \text{Mart}_c^{loc}$, there is a P-almost surely unique non-decreasing, P-almost surely continuous, progressively measurable function $\langle X\rangle$ such that $\langle X\rangle(0) \equiv 0$ and $X^2 - \langle X\rangle \in \text{Mart}_c^{loc}$. The uniqueness is an easy consequence of Corollary (2.24) (cf. part ii) of exercise (3.9) below). To prove existence, simply take $\langle X\rangle(t) = \sup_n \langle X^{\sigma_n}\rangle(t)$, $t \geq 0$. Finally, given $X, Y \in \text{Mart}_c^{loc}$, $\langle X,Y\rangle \equiv 1/4(\langle X+Y\rangle - \langle X-Y\rangle)$ is the P-almost surely unique progressively measurable function of local bounded variation which is P-almost surely continuous and satisfies $\langle X,Y\rangle(0) \equiv 0$ and $XY - \langle X,Y\rangle \in \text{Mart}_c^{loc}$.

(3.9) Exercise:

i) Let $X: [0,\infty)\times E \longrightarrow \mathbb{R}^1$ be a right continuous P-almost surely continuous progressively measurable function. Show that $(X(t), \mathcal{F}_t, P)$ is a martingale ($X \in \text{Mart}_c^2$) if and only if $X \in \text{Mart}_c^{loc}$ and there is a non-decreasing sequence of stopping times τ_n such that $\tau_n \longrightarrow \infty$ (a.s.,P) and $\{X(t\wedge\tau_n): n \geq 1\}$ is uniformly P-integrable ($\sup_{n\geq 1} E^P[X(t\wedge\tau_n)^2] < \infty$) for each $t \geq 0$.

ii) Show that if $X \in Mart_c^{loc}$ and $\zeta = \sup\{t\geq 0: |X|(t) < \infty\}$, then $X(t\wedge\zeta) = X(0)$, $t \geq 0$, (a.s.,P).

iii) Let $X \in Mart_c^{loc}$ and let $\alpha: [0,\infty)\times E \longrightarrow \mathbb{R}^1$ be a progressively measurable function satisfying $\int_0^T \alpha(t)^2\langle X\rangle(dt) < \infty$ (a.s.,P) for all $T \geq 0$. Show that there exists a $\int_0^\cdot \alpha(s)dX(s) \in Mart_c^{loc}$ such that $\langle\int_0^\cdot \alpha(s)dX(s),Y\rangle = \alpha\langle X,Y\rangle$ for all $Y \in Mart_c^{loc}$ and that, up to a P-null set, there is only one such element of $Mart_c^{loc}$. The quantity $\underline{\int_0^\cdot \alpha(s)dX(s)}$ is again called the <u>(Itô) stochastic integral of α with respect to X</u>.

iv) Suppose that $X \in (Mart_c^{loc})^M$ and that $Y: [0,\infty)\times E \longrightarrow \mathbb{R}^N$ is a right-continuous P-almost surely continuous progressively measurable function of local bounded variation. Set $Z = (X,Y)$ and let $f \in C^{2,1}(\mathbb{R}^M,\mathbb{R}^N)$ be given. Show that all the quantities in (3.6) are still well-defined and that (3.6) continues to hold. We will continue to refer to this extension of (3.6) as <u>Itô's formula</u>.

(3.10) <u>Lemma</u>: Let $X \in Mart_c^{loc}$ and let $\sigma \leq \tau$ be stopping times such that $\langle X\rangle(\tau) - \langle X\rangle(\sigma) \leq A$ for some $A < \infty$. Then,

$$P(\sup_{\sigma\leq t\leq\tau}|X(t)-X(\sigma)|\geq R) \leq 2\exp(-R^2/2A). \tag{3.11}$$

In particular, there exists for each $q \in (0,\infty)$ a universal $C_q < \infty$ such that

$$E^P\left[\sup_{\sigma\leq t\leq\tau}|X(t)-X(\sigma)|^q\right] \leq C_q A^{q/2}. \tag{3.12}$$

Proof: By replacing X with $X_\sigma = X - X^\sigma$, we see that it suffices to treat the case when $X(0) \equiv 0$, $\sigma \equiv 0$ and $\tau \equiv \infty$.

For $n \geq 1$, define $\zeta_n = \inf\{t\geq 0: \sup_{0\leq s\leq t}|X(s)|\geq n\}$ and set $X^n = X^{\zeta_n}$ and $Y_\lambda^n = \exp(\lambda X^n - \frac{\lambda^2}{2}\langle X^n\rangle)$. Then, by Itô's formula, $Y_\lambda^n(\cdot) = 1 + \lambda\int_0^\cdot Y_\lambda^n(s)dX^n(s) \in \mathrm{Mart}_c^2$. Hence, by Doob's inequality:

$$P(\sup_{0\leq t\leq T}X^n(t)\geq R) \leq P(\sup_{0\leq t\leq T}Y_\lambda^n(t)\geq \exp(\lambda R-\lambda^2 A/2))$$
$$\leq \exp(-\lambda R+\lambda^2 A/2)$$

for all $T > 0$ and $\lambda > 0$. After minimizing the right hand side with respect to $\lambda > 0$, letting n and T tend to ∞, and then repeating the argument with $-X$ replacing X, we obtain the required estimate. Clearly, (3.12) is an immediate consequence of (3.11).

Q.E.D.

(3.13) Exercise:

i) Suppose that $X \in (\mathrm{Mart}_c^{loc})^M$ and that τ is a stopping time for which $\sum_{i=1}^M \langle X^i\rangle(\tau) \leq A$ (a.s.,P) for some $A < \infty$. Let $\alpha: [0,\infty)\times E \longrightarrow \mathbb{R}^M$ be a progressively measurable function with the property that $\sup_{0\leq s\leq t}|\alpha(s)| \leq B\exp\left[\sup_{0\leq s\leq t}|X(s)-X(0)|^\gamma\right]$ (a.s.,P) for each $t \geq 0$ and some $\gamma \in [0,2)$ and $B < \infty$. Show that $(\sum_{i=1}^M\int_0^{t\wedge\tau}\alpha^i(s)dX^i(s),\mathcal{F}_t,P)$ is a martingale which is $L^q(P)$-bounded for every $q \in [1,\infty)$. In particular, show that if $\sum_{i=1}^M \langle X^i\rangle(T)$ is P-almost surely bounded for each $T \geq 0$ and if $f \in C^{2,1}(\mathbb{R}^M \times \mathbb{R}^N)$ satisfies the estimate $\max_{1\le i\le M} |\partial_{x^i} f(x,y)| \le A\exp(B|x|^\gamma)$, $(x,y) \in \mathbb{R}^M \times \mathbb{R}^N$, for some $A,B \in (0,\infty)$, then the stochastic integrals occuring in Itô's formula are elements of Mart_c^2.

ii) Given $X \in \mathrm{Mart}_c^{loc}$, set $\mathcal{E}_X(t) = \exp[X(t)-1/2\langle X\rangle(t)]$ for $t \ge 0$. Show that $\mathcal{E}_X$ is the P-almost surely unique $Y \in \mathrm{Mart}_c^{loc}$ such that $Y(t) = 1 + \int_0^t Y(s)dX(s)$, $t \ge 0$, (a.s.,P). (Hint: To prove uniqueness, consider $Y(t)/\mathcal{E}_X(t)$.) Also, if $\|\langle X\rangle(\tau)\|_{L^\infty(P)} < \infty$ for some finite stopping time τ, show that $(\mathcal{E}_X(t\wedge\tau,\mathcal{F}_t,P)$ is a martingale and that, for each $q \in [1,\infty)$, $\|\mathcal{E}_X(\tau)\|_{L^q(P)} \le \exp[(q-1)\|\langle X\rangle(\tau)\|_{L^\infty(P)}]$. The quantity $\mathcal{E}_X$ is sometimes called the <u>Itô exponential of X</u>.

The following exercise contains a discussion of an important and often useful representation theorem. Loosely speaking, what it says is that an $X \in \mathrm{Mart}_c^{loc}$ "has the paths of a Brownian motion and uses $\langle X\rangle$ as its clock."

(3.14) <u>Exercise</u>:

i) Let $(X(t),\mathcal{F}_t,P)$ be a martingale and for each $t \ge 0$ let $\overline{\mathcal{F}}_{t+}$ denote the P-completion of $\mathcal{F}_{t+} \equiv \bigcap_{\epsilon>0} \mathcal{F}_{t+\epsilon}$. Show that $(X(t),\overline{\mathcal{F}}_{t+},P)$ is again a martingale.

ii) Let $Y: [0,\infty)\times E \longrightarrow \mathbb{R}^1$ be a measurable function such that $t \longmapsto Y(t,\xi)$ is right-continuous for every $\xi \in E$. Assuming that for each $T > 0$:

$$\sup_{0\le s<t\le T} E^P[|Y(t)-Y(s)|^q]/(t-s)^{1+\alpha} < \infty,$$

show that $t\longmapsto Y(t,\xi)$ is continuous for P-almost every $\xi \in E$. (Hints: For $n \geq 0$ let

$$Y_n(t) = Y(k/2^n) + 2^n(t-k/2^n)(Y((k+1)/2^n) - Y(k/2^n))$$

for $t \in [k/2^n,(k+1)/2^n)$ and $k \geq 0$. Show that for any $\beta \in (0,1]$ and $T \in Z^+$

$$\sup_{0\le s<t\le T} \frac{|Y_n(t)-Y_n(s)|}{(t-s)^\beta} \le \sup_{0\le s<t\le T} \frac{|Y_{n+1}(t)-Y_{n+1}(s)|}{(t-s)^\beta}.$$

Next, using Theorem (I.2.12), show that for each $T \in Z^+$ there is a $\beta \in (0,1]$ such that:

$$\lim_{R\longrightarrow\infty} \sup_{n\ge 0} P\left[\sup_{0\le s<t\le T} \frac{|Y_n(t)-Y_n(s)|}{(t-s)^\beta} \ge R\right] = 0;$$

and conclude that

$$\lim_{R\longrightarrow\infty} P\left[\sup_{n\ge 0}\ \sup_{0\le s<t\le T} \frac{|Y_n(t)-Y_n(s)|}{(t-s)^\beta} \ge R\right] = 0.)$$

iii) Let $X \in Mart_c^{loc}(\{\mathcal{F}_t\},P)$ and define $\tau(t) = \sup\{s\ge 0: \langle X\rangle(s) \le t\}$. Show that $t\longmapsto\tau(t)$ is right-continuous and non-decreasing and that, for each $t \geq 0$, $\tau(t)$ is an $\{\overline{\mathcal{F}}_{s+}\}$-stopping time. Next, set $\mathcal{G}_t = \overline{\mathcal{F}}_{\tau(t)+}$ ($\equiv \{A: A\cap\{\tau(t)\le s\} \in \overline{\mathcal{F}}_{s+}$ for all $s \geq 0\}$); and define $Z: [0,\infty)\times E\longrightarrow\mathbb{R}^1$ so that:

$$Z(t) = \begin{matrix} X(\tau(t)) - X(0) \text{ if } \tau(t) < \infty \\ X(\infty) - X(0) \text{ if } \tau(t) = \infty \end{matrix}$$

where $X(\infty) = \lim_{s\longrightarrow\infty} X(s)$ when this limit exists in $(-\infty,\infty)$ and $X(0)$ otherwise. Show that: $\langle X\rangle(s)$ is a $\{\mathcal{G}_t\}$-stopping time for each $s \geq 0$, $Z(0) = 0$ (a.s.,P), $(Z(t),\mathcal{G}_t,P)$ is a martingale, and that $E^P[|Z(t)-Z(s)|^4] \le C_4(t-s)^2$, $0 \le s < t < \infty$ (where C_4 is the constant in (3.12)). Conclude that $Z \in Mart_c^2(\{\mathcal{G}_t\},P)$ and show that $\langle Z\rangle(dt) \le dt$ (a.s.,P). Finally, show that, for each $T > 0$, $\langle Z\rangle(t) = t$, $t \in [0,T]$, (a.s.,P) on $\{\langle X\rangle(\infty) > T\}$. In particular, if $\langle X\rangle(\infty) = \infty$ (a.s.,P), set $B = Z - Z(0)$ and conclude that $(B(t),\mathcal{G}_t,P)$ is a one-dimensional Brownian motion and that

$$X(t) = X(0) + B(\langle X\rangle(t)),\ t \geq 0,\ (\text{a.s.},P). \tag{3.15}$$

iv) To extend the representation in (3.15) to cases in which $\langle X\rangle = \infty$ (a.s.,P) may fail, proceed as follows. Let $\mathcal{W}$ denote one-dimensional Wiener measure on $(\Omega,\mathcal{M})$ and set $Q = P\times\mathcal{W}$ and $\mathcal{H}_t = \mathcal{G}_t\times\mathcal{M}_t$, $t \geq 0$. Let $Z(\cdot)$ be as in iii); but now define $B\colon [0,\infty)\times E\times\Omega \longrightarrow \mathbb{R}^1$ by:

$$B(t,(\xi,\omega)) = Z(t\wedge\langle X\rangle(\infty,\xi),\xi)$$
$$- Z(0) + x(t,\omega) - x(t\wedge\langle X\rangle(\infty,\xi),\omega).$$

Show that $(B(t),\mathcal{H}_t,Q)$ is a one-dimensional Brownian motion and that (3.15) holds (a.s.,Q).

As an application of the preceding, consider the following.

(3.16) <u>Exercise</u>: Let $X \in \text{Mart}_c^{loc}$. Show that $\{\lim_{s\to\infty} X(s)$ exists in $(-\infty,\infty)\} = \{\lim_{s\to\infty} X(s)$ exists in $[-\infty,\infty]\} = \{\langle X\rangle(\infty) < \infty\}$ (a.s.,P). (Hint: Prove that if $\beta(\cdot)$ is a one-dimensional Brownian motion, then $\overline{\lim}_{s\to\infty}\beta(s) = -\underline{\lim}_{s\to\infty}\beta(s) = \infty$ almost surely.)

<u>Remark</u>: As a consequence of (3.16), we see that there is no hope of defining $\int_0^T \alpha(s)dX(s)$ for α's which fail to satisfy $\int_0^T \alpha(s)^2\langle X\rangle(ds) < \infty$.

We have seen in Lemma (3.10) that $E^P[\sup_{0\le t\le T}|X(t)-X(0)|^q] \le C_q \|\langle X\rangle(T)\|^{q/2}_{L^\infty(P)}$ for $X \in \text{Mart}_c^{\text{loc}}$. At least when $q \in [2,\infty)$, we are now going to prove a refinement of this result. The inequalities which we have in mind are referred to as Burkholder's inequality; however, the proof which we are about to give is due to A. Garsia and takes full advantage of the fact that we are dealing with continuous martingales.

(3.17) Theorem (Burkholder's Inequality): For each $q \in [2,\infty)$, all $X \in \text{Mart}_c^{\text{loc}}$, and all stopping times τ:

$$
\begin{aligned}
a_q \|\langle X\rangle(\tau)^{1/2}\|_{L^q(P)} &\le \|(X-X(0))^*(\tau)\|_{L^q(P)} \\
&\le A_q \|\langle X\rangle(\tau)^{1/2}\|_{L^q(P)}
\end{aligned}
\tag{3.18}
$$

where $a_q = (2q)^{-1/2}$ and $A_q = 2^{1/2}q(q')^{(q-1)/2}$ $(1/q' = 1 - 1/q)$.

Proof: First note that it suffices to prove (3.18) when $X(0) \equiv 0$ and τ, $X(\tau)$, and $\langle X\rangle(\tau)$ are all bounded. Second, by replacing X with X^τ if necessary, we can reduce to the case when $\tau \equiv T < \infty$ and X and $\langle X\rangle$ are bounded. Hence, we will prove (3.18) under these conditions. In particular, this means that $X \in \text{Mart}_c^2$.

To prove the right hand side, apply Itô's formula to write:

$$
\begin{aligned}
|X(T)|^q = q\int_0^T \text{sgn}(X(t))\,|X(t)|^{q-1}dX(t) \\
+ \frac{q(q-1)}{2}\int_0^T |X(t)|^{q-2}\langle X\rangle(dt).
\end{aligned}
$$

Then, by (2.11):

$$(1/q')^q E^P[X^*(T)^q] \le E^P[|X(T)|^q]$$

$$= E^P\left[\frac{q(q-1)}{2}\int_0^T |X(t)|^{q-2}\langle X\rangle(dt)\right]$$

$$\le \frac{q(q-1)}{2}E^P[X^*(T)^{q-2}\langle X\rangle(T)]$$

$$\le \frac{q(q-1)}{2}E^P[X^*(T)^q]^{1-2/q}E^P[\langle X\rangle(T)^{q/2}]^{1/q},$$

from which the right hand side of (3.18) is immediate.

To prove the left hand side of (3.18), note that, by Itô's formula:

$$X(T)\langle X\rangle(T)^{(q-2)/4} = Y(T) + \int_0^T X(t)\langle X\rangle^{(q-2)/4}(dt),$$

where $Y(\cdot) \equiv \int_0^{\cdot} \langle X\rangle(t)^{(q-2)/4}dX(t)$. Hence:

$$|Y(T)| \le 2X^*(T)\langle X\rangle(T)^{(q-2)/4}.$$

At the same time:

$$\langle Y\rangle(T) = \int_0^T \langle X\rangle(t)^{(q-2)/2}\langle X\rangle(dt) = \frac{2}{q}\langle X\rangle(T)^{q/2}.$$

Thus:

$$E^P[\langle X\rangle(T)^{q/2}] = \frac{q}{2}E^P[\langle Y\rangle(T)] = \frac{q}{2}E^P[Y(T)^2]$$

$$\le 2qE^P[X^*(T)^2\langle X\rangle(T)^{(q-2)/2}]$$

$$\le 2qE^P[X^*(T)^q]^{2/q}E^P[\langle X\rangle(T)^{q/2}]^{1-2/q}.$$

Q.E.D.

<u>Remark</u>: It turns out that (3.18) actually holds for all $q \in (0,\infty)$ with appropriate choices of a_q and A_q. When $q \in (1,2]$, this is again a result due to D. Burkholder; for $q \in (0,1]$, it was first proved by D. Burkholder and R. Gundy using a quite intricate argument. However, for continuous

martingales, A. Garsia showed that the proof for $q \in (0,2]$ can be again greatly simplified by clever application of Itô's formula (cf. Theorem 3.1 in <u>Stochastic Differential Equations and Diffusion Processes</u> by N. Ikeda and S. Watanabe, North Holland, 1981).)

Before returning to our main line of development, we will take up a particularly beautiful application of Itô's formula to the study of Brownian paths.

(3.19) <u>Theorem</u>: Let $(\beta(t),\mathscr{F}_t,P)$ be a one-dimensional Brownian motion and assume that the $\mathscr{F}_t$'s are P-complete. Then there exists a P-almost surely unique function ℓ: $[0,\infty)\times\mathbb{R}^1\times E\longrightarrow[0,\infty)$ such that:

i) For each $x \in \mathbb{R}^1$, $(t,\xi)\longmapsto\ell(t,x,\xi)$ is progressively measurable; for each $\xi \in E$, $(t,x)\longmapsto\ell(t,x,\xi)$ is continuous; and, for each $(x,\xi) \in \mathbb{R}^1\times E$, $\ell(0,x,\xi) = 0$ and $t\longmapsto\ell(t,x,\xi)$ is non-decreasing.

ii) For all bounded measurable φ: $\mathbb{R}^1\longrightarrow\mathbb{R}^1$:

$$\int_{\mathbb{R}^1}\varphi(y)\ell(t,y)dy = 1/2\int_0^t\varphi(\beta(s))ds,\ t \geq 0,\ (\text{a.s.},P). \tag{3.20}$$

Moreover, for each $y \in \mathbb{R}^1$:

$$\ell(t,y) = \beta(t)\vee 0 - \int_0^t\chi_{[y,\infty)}(\beta(s))d\beta(s),\ t \geq 0, \tag{3.21}$$

(a.s.,P).

<u>Proof</u>: Clearly i) and ii) uniquely determine ℓ. To see how one might proceed to construct ℓ, note that (3.20) can be interpreted as the statement that "$\ell(t,y) =$

$1/2\int_0^t \delta(\beta(s)-y)ds$", where δ is the Dirac δ-function. This interpretation explains the origin of (3.21). Indeed, $(\cdot \vee y)'$ $= \chi_{[y,\infty)}(\cdot)$ and $(\cdot \vee y)'' = \delta(\cdot - y)$. Hence, (3.21) is precisely the expression for ℓ predicted by Itô's formula. In order to justify this line of reasoning, it will be necessary to prove that there is a version of the right hand side of (3.21) which has the properties demanded by i).

To begin with, for fixed y, let $t \longmapsto k(t,y)$ be the right hand side of (3.21). We will first check that $k(\cdot,y)$ is P-almost surely non-decreasing. To this end, choose $\rho \in C_0^\infty(\mathbb{R}^1)^+$ having integral 1, and define $f_n(x) =$ $n\int \rho(n(x-\zeta))(\zeta \vee y)d\zeta$ for $n \geq 1$. Then, by Itô's formula:

$$f_n(\beta(t)) - f_n(0) - \int_0^t f_n'(\beta(s))d\beta(s) = 1/2\int_0^t f_n''(\beta(s))ds$$

(a.s.,P). Because $f_n'' \geq 0$, we conclude that the left hand side of the preceding is P-almost surely non-decreasing as a function of t. In addition, an easy calculation shows that the left hand side tends, P-almost surely, to $k(\cdot,y)$ uniformly on finite intervals. Thus $k(\cdot,y)$ is P-almost surely non-decreasing.

We next show that, for each y, $k(\cdot,y)$ can be modified on a set of P-measure 0 in such a way that the modified function is continuous with respect to (t,y). Using (I.2.12) in the same way as was suggested in the hint for part ii) of (3.14), one sees that this reduces to checking that:

$$E^P\left[\sup_{0\leq t\leq T}\left|\int_0^t \chi_{[y,\infty)}(\beta(s))d\beta(s) - \int_0^t \chi_{[x,\infty)}(\beta(s))d\beta(s)\right|^4\right] \leq CT(y-x)^2$$

for some $C < \infty$ and all $(T,x,y) \in [0,\infty)\times\mathbb{R}^2$. But, by (2.11), this comes down to estimating $E^P\left[\left[\int_0^T \chi_{[x,y)}(\beta(s))d\beta(s)\right]^4\right]$ for $x < y$; and, by (3.18), this, in turn, reduces to estimating $E^P\left[\left[\int_0^T \chi_{[x,y)}(\beta(s))ds\right]^2\right]$. But

$$E^P\left[\left[\int_0^T \chi_{[x,y)}(\beta(s))ds\right]^2\right]$$

$$= 2E^P\left[\int_0^T \chi_{[x,y)}(\beta(t))dt\int_0^t \chi_{[x,y)}(\beta(s))ds\right]$$

$$= 2\int_0^T dt\int_0^t ds\int_x^y d\zeta g(s,\zeta)\int_x^y g(t-s,\eta-\zeta)d\eta,$$

where g is the one-dimensional Gauss kernel, and the required estimate is immediate from here.

We now know that there is an ℓ which satisfies both i) and (3.21). To prove that it also satisfies (3.20), set $M(t,y) = \beta(t)\vee y - 0\vee y - \ell(t,y)$. Then

$$M(\cdot,y) = \int_0^{\cdot} \chi_{[y,\infty)}(\beta(s))d\beta(s) \quad (\text{a.s.},P)$$

for each y. Hence, if $\varphi \in C_0(\mathbb{R}^1)$ and we use Riemann approximations to compute $\int \varphi(y)M(t,y)dy$, then it is clear that $\int \varphi(y)M(\cdot,y)dy \in \text{Mart}_c^{loc}$. In particular, we now see that

$$\Phi(\beta(\cdot)) - \Phi(\beta(0)) - \int \varphi(y)M(\cdot,y)dy \in \text{Mart}_c^{loc},$$

where $\Phi(x) = \int (x\vee y)\varphi(y)dy$. On the other hand, by Itô's formula:

$$\Phi(\beta(\cdot)) - \Phi(\beta(0)) - 1/2\int_0^{\cdot} \varphi(\beta(s))ds \in \text{Mart}_c^{loc}.$$

Thus, by part ii) of (3.9), $\int\varphi(y)\ell(\cdot,y)dy = 1/2\int_0^{\cdot}\varphi(\beta(s))ds$ (a.s.,P); and clearly (3.20) follows from this. Q.E.D.

Remark: The function $\ell(\cdot,y)$ described above is called the local time of β at y. ℓ was first discussed by P. Levy and its existence was first proved by H. Trotter. The beautiful and simple development given above is the idea of H. Tanaka, and (3.21) is sometimes referred to as Tanaka's formula.

(3.22) Exercise: The notation is the same as that in Theorem (3.19).

i) Show that:

$$|\beta(t)| = \int_0^t \mathrm{sgn}(\beta(s))d\beta(s) + 2\ell(t,0),\ t \geq 0, \tag{3.23}$$

(a.s.,P).

ii) Show that if $A \in \mathcal{M}_{0+}$ $(\equiv \bigcap_{\epsilon>0}\mathcal{M}_\epsilon)$, then $\mathcal{W}(A) \in \{0,1\}$. (Hint: Note that, for any $\Phi \in C_b(\Omega)$ and $t > 0$, $E^{\mathcal{W}}[\Phi\circ\theta_t, A] = E^{\mathcal{W}}[E^{\mathcal{W}_{x(t)}}[\Phi], A]$. Let $t\downarrow 0$ and conclude that $E^{\mathcal{W}}[\Phi, A] = E^{\mathcal{W}}[\Phi]\mathcal{W}(A)$ for all $\Phi \in C_b(\Omega)$ and therefore that A is independent of $\mathcal{M}$.)

iii) Set $B(\cdot) = \int_0^{\cdot}\mathrm{sgn}(\beta(s))d\beta(s)$ and note that $(B(t),\mathcal{F}_t,P)$ is a one-dimensional Brownian motion. Using ii), show that $P(\inf_{0<t<\delta}B(t)<0 \text{ for all } \delta>0) = 1$; and conclude from this and (3.23) that $P(\ell(t,0)>0 \text{ for all } t>0) = 1$.

iv) Show that, for each y, $\ell(\{t\geq 0:\ \beta(t,\xi)\neq y\},y,\xi) = 0$ for P-almost every ξ, and conclude that $\ell(dt,y,\xi)$ is singular with respect to dt for P-almost every ξ. In particular, $\ell(\cdot,0)$ is P-almost surely a continuous, non-decreasing function on $[0,\infty)$ such that $\ell(dt,0)$ is singular with respect to dt and $\ell(t,0) > 0$ for all $t > 0$.

We at last pick up the thread which we dropped after introducing the class $Mart_c^{loc}$. Recall that we were attempting to find a good description of the class of processes generated by $Mart_c^2$ under changes of coordinates. We are now ready to give that description. Denote by $\underline{B}.\underline{V}._c$ the class of right continuous, P-almost surely continuous, progressively measurable $Y: [0,\infty)\times E \longrightarrow \mathbb{R}^1$ which are of local bounded variation. We will say that $(Z(t),\mathcal{F}_t,P)$ is a P-almost surely continuous semi-martingale and will write $Z \in$ $\underline{S}.\underline{Mart}_c$ if Z can be written as the sum of a martingale part X $\in Mart_c^{loc}$ and a locally bounded variation part $Y \in B.V._c$. Note that, up to a P-null set, the martingale part X and the locally bounded variation part Y of a $Z \in S.Mart_c$ are uniquely determined (cf. part ii) of (3.9)). Moreover, by Itô's formula, if $Z \in (S.Mart_c)^N$ and $f \in C^2(R^N)$, then $f\circ Z \in S.Mart_c$. Thus $S.Mart_c$ is certainly invariant under changes of coordinates.

Given $Z \in S.Mart_c$ with martingale part X and locally bounded variation part Y, we will use $\langle Z\rangle$ to denote $\langle X\rangle$; and if Z' is a second element of $S.Mart_c$ with associated parts X' and Y', we use $\langle Z,Z'\rangle$ to denote $\langle X,X'\rangle$. Also, if α: $[0,\infty)\times E \longrightarrow \mathbb{R}^1$ is a progressively measurable function satisfying

$$\int_0^T \alpha(t)^2\langle Z\rangle(dt) \vee \int_0^T |\alpha(t)|\,|Y|(dt) < \infty,\ T > 0, \tag{3.24}$$

(a.s.,P), we define $\int_0^{\cdot} \alpha(s)dZ(s)$ to be $\int_0^{\cdot} \alpha(s)dX(s) + \int_0^{\cdot} \alpha(s)Y(ds)$. Notice that in this notation, Itô's formula for P-almost surely continuous semi-martingales becomes

$$\begin{aligned} f(Z(t)) - f(Z(0)) = &\sum_{i=1}^{N} \int_0^t \partial_{z^i} f(Z(s))dZ^i(s) \\ &+ 1/2 \sum_{i,j=1}^{N} \int_0^t \partial_{z^i}\partial_{z^j} f(Z(s))\langle Z^i, Z^j\rangle(ds) \end{aligned} \tag{3.25}$$

for $Z \in (S.Mart_c)^N$ and $f \in C^2(\mathbb{R}^N)$.

(3.26) Exercise: Let $Z \in (S.Mart_c)^N$ and $f \in C^2(\mathbb{R}^N)$. Show that, for any $Y \in S.Mart_c$, $\langle f\circ Z, Y\rangle = \sum_{i=1}^{N} (\partial_{z^i} f)\circ Z\langle Z^i, Y\rangle$ (a.s.,P).

We conclude this section with a brief discussion of the Stratonovich integral as interpreted by Itô. Namely, given $X, Y \in S.Mart_c$, define the Stratonovich integral $\int_0^{\cdot} X(s)\circ dY(s)$ of X with respect to Y (the "$\circ$" in front of the dY(s) is put there to emphasize that this is not an Itô integral) to be the element of $S.Mart_c$ given by $\int_0^{\cdot} X(s)dY(s) + 1/2\langle X, Y\rangle(\cdot)$. Although the Stratonovich integral appears to be little more than a strange exercise in notation, it turns out to be a very useful device. The origin of all its virtues is contained in the form which Itô's formula takes when Stratonovich integrals are used. Namely, from (3.25) and (3.26), we see that Itô's formula becomes the fundamental theorem of calculus:

$$f(Z(t)) - f(Z(0)) = \sum_{i=1}^{N} \int_0^t \partial_{z^i} f(Z(s)) \circ dZ(s) \qquad (3.27)$$

for all $Z \in (S.Mart_c)^N$ and $f \in C^3(R^N)$. The major drawback to the Stratonovich integral is that it requires that the integrand be a semimartingale (this is the reason why we restricted f to lie in $C^3(\mathbb{R}^N)$). However, in some circumstances, Itô has shown how even this drawback can be overcome.

(3.28) <u>Exercise</u>: Given $X, Y, Z \in S.Mart_c$, show that:

$$\int_0^{\cdot} X(t) \circ d\left[\int_0^t Y(s) \circ dZ(s)\right] = \int_0^{\cdot} (XY)(s) \circ dZ(s). \qquad (3.29)$$

III. THE MARTINGALE PROBLEM FORMULATION OF DIFFUSION THEORY:

1. Formulation and Some Basic Facts:

Recall the notation Ω, $\mathcal{M}$, and $\{\mathcal{M}_t\}$ introduced at the beginning of section I.3.

Given bounded measurable functions $a\colon [0,\infty)\times R^N \longrightarrow S^+(\mathbb{R}^N)$ (cf. the second paragraph of (II.1)) and $b\colon [0,\infty)\times\mathbb{R}^N \longrightarrow \mathbb{R}^N$, define $t \longmapsto L_t$ by (II.1.1). Motivated by the results in Corollary (II.1.13), we now pose the <u>martingale problem for</u> $\{L_t\}$. Namely, we say that $P \in \mathbf{M}_1(\Omega)$ <u>solves the martingale problem for</u> $\{L_t\}$ <u>starting from</u> $(s,x) \in [0,\infty)\times\mathbb{R}^N$ <u>and write</u> $P \in$ $\mathbf{M}.\mathbf{P}.((s,x);\{L_t\})$ if:

$$
\begin{aligned}
&\text{i)}\ P(x(0)=x) = 1, \\
&\text{ii)}\ \left(\varphi(x(t))-\int_0^t [L_{s+u}\varphi](x(u))du, \mathcal{M}_t, P\right) \text{ is a martingale} \\
&\qquad \text{for every } \varphi \in C_0^\infty(\mathbb{R}^N).
\end{aligned} \tag{M.P.}
$$

Given $\varphi \in C^2(\mathbb{R}^N)$, set

$$X_{s,\varphi}(t) = \varphi(x(t)) - \int_0^t [L_{s+u}\varphi](x(u))du. \tag{1.1}$$

If $P \in M.P.((s,x);\{L_t\})$, then $X_{s,\varphi} \in Mart_c^2(\{\mathcal{M}_t\},P)$ for every $\varphi \in C_0^\infty(R^N)$. To compute $\langle X_{s,\varphi}\rangle$, note that:

$$X_{s,\varphi}(t)^2 = \varphi(x(t))^2 - 2\varphi(x(t))\int_0^t [L_{s+u}\varphi](x(u))du + \left[\int_0^t [L_{s+u}\varphi](x(u))du\right]^2$$

$$= \varphi(x(t))^2 - 2X_{s,\varphi}(t)\int_0^t [L_{s+u}\varphi](x(u))du - \left[\int_0^t [L_{s+u}\varphi](x(u))du\right]^2$$

$$= X_{s,\varphi^2}(t) + \int_0^t [L_{s+u}\varphi^2](x(u))du - 2X_{s,\varphi}(t)\int_0^t [L_{s+u}\varphi](x(u))du - \left[\int_0^t [L_{s+u}\varphi](x(u))du\right]^2.$$

Applying Lemma (II.2.22) (or Itô's formula), we see that:

$$\left(X_{s,\varphi}(t)^2 - \int_0^t \left[[L_{s+u}\varphi^2](x(u)) - 2[\varphi L_{s+u}\varphi](x(u))\right]du, \mathcal{M}_t, P\right)$$

is a martingale. Noting that $[L_{s+u}\varphi^2](y) - 2[\varphi L_{s+u}\varphi](y) = (\nabla\varphi, \mathbf{a}\nabla\varphi)(s+u,y)$, we conclude that:

$$\langle X_{s,\varphi}\rangle(\cdot) = \int_0^{\cdot} (\nabla\varphi, \mathbf{a}\nabla\varphi)(s+u, x(u))du, \text{ (a.s., P)}. \tag{1.2}$$

<u>Remark</u>: As an immediate consequence of (1.2), we see that when $\mathbf{a} \equiv 0$, $X_{s,\varphi}(t) = X_{s,\varphi}(0) = \varphi(x)$, $t \geq 0$, (a.s.,P) for each $\varphi \in C_0^\infty(\mathbb{R}^N)$. From this it is clear that:

$$x(t,\omega) = x + \int_0^t b(s+u, x(u,\omega))du, \ t \geq 0,$$

for P-almost every $\omega \in \Omega$. In other words, when $\mathbf{a} \equiv 0$ and $P \in M.P.((s,x);\{L_t\})$, P-almost every $\omega \in \Omega$ is an integral curve of the time dependent vector field $b(\cdot,*)$ starting from (s,x). This is, of course, precisely what we would expect on the basis of (II.1.16) and (II.1.17).

Again suppose that $P \in M.P.((s,x);\{L_t\})$. Given $\varphi \in C^2(\mathbb{R}^N)$, note that the quantity $X_{s,\varphi}$ in (1.1) is an element of $Mart_c^{loc}(\{\mathcal{M}_t\},P)$. Indeed, by an easy approximation procedure, it is clear that $X_{s,\varphi} \in Mart_c^2(\{\mathcal{M}_t\},P)$ when $\varphi \in C_b^2(\mathbb{R}^N)$. For general $\varphi \in C^2(\mathbb{R}^N)$, set $\sigma_n = \inf\{t\ge 0: |x(t)|\ge n\}$ and choose $\eta_n \in C_0^\infty(\mathbb{R}^N)$ so that $\eta_n(y) = 1$ for $|y| \le (n+1)$, $n \ge 1$. Then, $\sigma_n\uparrow\infty$ and $X_{s,\varphi}(\cdot\wedge\sigma_n) = X_{s,\eta_n\varphi}(\cdot\wedge\sigma_n) \in \mathrm{Mart}_c^2(\{\mathcal{M}_t\},P)$. At the same time, it is clear that (1.2) continues to hold for all $\varphi \in C^2(\mathbb{R}^N)$. In particular, if:

$$\bar{x}(t) = x(t) - \int_0^t b(s+u,x(u))du, \quad t \ge 0, \tag{1.3}$$

then $\bar{x} \in (\mathrm{Mart}_c^{loc}(\{\mathcal{M}_t\},P))^N$ and

$$\langle\langle\bar{x},\bar{x}\rangle\rangle(t) = \int_0^t a(s+u,x(u))du, \quad t \ge 0, \ (\text{a.s.},P). \tag{1.4}$$

Using (1.4) and applying Lemma (II.3.10) (cf. the proof of (II.1.6) as well), we now see that:

$$\begin{aligned} P(\sup_{s\le t\le T}|\bar{x}(t)-\bar{x}(s)| \ge R) \\ \le 2N\exp[-R^2/2AN(T-s)], \quad 0 \le s < T, \end{aligned} \tag{1.5}$$

where $A \equiv \sup_{t,y} \|a(t,y)\|_{op}$. From (1.5) it follows that for each $q \in (0,\infty)$ there is a universal $C(q,N) \in [1,\infty)$ such that:

$$\left\|\sup_{s\le t\le T}|\bar{x}(t)-\bar{x}(s)|\right\|_{L^q(P)} \le C(q,N)(A(T-s))^{q/2}, \tag{1.6}$$

$0 \le s < T$. In particular, this certainly means that $\bar{x} \in (\mathrm{Mart}_c^2(\{\mathcal{M}_t\},P))^N$. In addition, by Itô's formula, for any $f \in C^{1,2}([0,\infty)\times\mathbb{R}^N)$, P-almost surely it is true that:

$$\begin{aligned} f(\cdot,x(\cdot)) - f(0,x) - \int_0^{\cdot}[(\partial_u+L_{s+u})f](u,x(u))du \\ = \int_0^{\cdot}[\nabla_x f](u,x(u))\cdot d\bar{x}(u) \end{aligned} \tag{1.7}$$

(where $\int_0^t[\nabla_x f](u,x(u))\cdot d\bar{x}(u) \equiv \sum_{i=1}^N\int_0^t \partial_{x^i} f(u,x(u))d\bar{x}^i(s)$); and, by part i) of exercise (II.3.13), if $|\nabla_y f(t,y)| \le$

$A\exp(B|y|^{\gamma})$, $(t,y) \in [0,\infty)$, for some $A,B \in [0,\infty)$ and $\gamma \in [0,2)$, then the right hand side of (1.7) is an element of $Mart_c^2(\{\mathcal{M}_t\},P)$.

(1.8) <u>Exercise</u>: Show that $P \in M.P.((s,x);\{L_t\})$ if and only if $P(x(0)=x) = 1$, $\bar{x} \in (Mart_c^{loc}(\{\mathcal{M}_t\},P))^N$, and (1.4) holds.

(1.9) <u>Exercise</u>: The considerations discussed thus far in this section can all be generalized as follows. Let $a: [0,\infty)\times\Omega \longrightarrow S^+(\mathbb{R}^N)$ and $b: [0,\infty)\times\mathbb{R}^N \longrightarrow \mathbb{R}^N$ be bounded $\{\mathcal{M}_t\}$-progressively measurable functions and define $t \longmapsto L_t$ by analogy with (II.1.1). Define $M.P.(x;\{L_t\})$ to be the set of $P \in \mathbf{M}_1(\Omega)$ such that $P(x(0)=x) =1$ and

$$\left(\varphi(x(t))-\int_0^t [L_u\varphi](x(u))du, \mathcal{M}_t, P\right)$$

is a martingale for all $\varphi \in C_0^{\infty}(\mathbb{R}^N)$. Define $\bar{x}(t) = x(t) - \int_0^t b(u)du$ and show that $P \in M.P.(x;\{L_t\})$ if and only if $P(x(0)=x) = 1$, $\bar{x} \in (Mart_c^{loc}(\{\mathcal{M}_t\},P))^N$ and $\langle\langle \bar{x},\bar{x}\rangle\rangle(\cdot) = \int_0^{\cdot} a(u)du$ (a.s.,P). Also, show that if $P \in M.P.(x;\{L_t\})$, then (1.5), (1.6), and (1.7) continue to hold and that the right hand side of (1.7) is an element of $Mart_c^2(\{\mathcal{M}_t\},P)$ under the same conditions as those given following (1.7).

(1.10) <u>Remark</u>: When a and b do not depend on $t \in [0,\infty)$, denote L_0 by L and note that $M.P.((s,x);\{L_t\})$ is independent of $s \in [0,\infty)$. Thus we will use <u>M.P.(x;L)</u> in place of M.P.$((s,x);\{L_t\})$ for time-independent coefficients a and b.

(1.11) <u>Exercise</u>: Time-dependent martingale problems can be thought of as time-independent ones via the following trick. Set $\tilde{\Omega} = C([0,\infty);\mathbb{R}^{N+1})$ and identify $C([0,\infty);\mathbb{R}^{N+1})$ with $C([0,\infty);\mathbb{R}^1)\times\Omega$. Show that $P \in$ M.P.$((s,x);\{L_t\})$ if and only if $\tilde{P} \in$ M.P.$(\tilde{x};\tilde{L})$, where $\tilde{x} = (s,x)$, $\tilde{L} = \partial_t + L_t$, $\tilde{P} = \delta_{s+\cdot}\times P$, and $\delta_{s+\cdot} \in M_1(C([0,\infty);\mathbb{R}^1)$ denotes the delta mass at the path $t \longmapsto s+t$.

The following result is somewhat technical and can be avoided in most practical situations. Nonetheless, it is of some theoretical interest and it smooths the presentation of the general theory.

(1.12) <u>Theorem</u>: Assume that for each $(s,x) \in [0,\infty)\times\mathbb{R}^N$ the set M.P $((s,x);\{L_t\})$ contains precisely one element $P_{s,x}$. Then, $(s,x) \longmapsto P_{s,x} \in M_1(\Omega)$ is a Borel measurable map.

<u>Proof</u>: In view of (1.11), we loose no generality by assuming that a and b are independent of time. Thus we do so, and we will show that $x \longmapsto P_x$ is measurable under the assumption that P_x is the unique element of M.P.(x;L) for each $x \in \mathbb{R}^N$.

Define Γ to be the subset of $M_1(\Omega)$ consisting of those P with the property that $(\varphi(x(t))-\int_0^t [L\varphi](x(u))du, \mathcal{M}_t, P)$ is a martingale for all $\varphi \in C_0^\infty(\mathbb{R}^N)$. Clearly, there is a sequence

$\{X_n\}$ of bounded measurable functions on Ω such that $P \in \Gamma$ if and only if $E^P[X_n] = 0$ for all $n \geq 1$. Thus Γ is a Borel measurable subset of $\mathbf{M}_1(\Omega)$. In the same way, one can show that the set Δ consisting of those $P \in \mathbf{M}_1(\Omega)$ such that $P\circ(x(0))^{-1} = \delta_x$ for some $x \in \mathbb{R}^N$ is a Borel subset of $\mathbf{M}_1(\Omega)$. Thus, $\Gamma_0 \equiv \cup\{M.P.(x;L): x \in R^N\} = \Gamma\cap\Delta$ is a Borel subset of $\mathbf{M}_1(\Omega)$. At the same time, $P \in \Gamma_0 \longmapsto P\circ(x(0))^{-1} \in \mathbf{M}_1(\mathbb{R}^N)$ is clearly a measurable mapping, and, by assumption, it is one-to-one. In particular, since $\mathbf{M}_1(\Omega)$ and $\mathbf{M}_1(\mathbb{R}^N)$ are Polish spaces (see (I.2.5)), the general theory of Borel mappings on Polish spaces says that the inverse map $\delta_x \longmapsto P_x$ is also a measurable mapping (cf. Theorem 3.9 in K. R. Parthasarathy's <u>Probability Measures on Metric Spaces</u>, Academic Press, 1967). But $x \longmapsto \delta_x$ is certainly measurable, and therefore we have now shown that $x \longmapsto P_x$ is measurable. Q.E.D.

(1.13) <u>Exercise</u>: Suppose that τ is an $\{\mathcal{M}_t\}$-stopping time. Show that $\mathcal{M}_\tau = \sigma(x(t\wedge\tau): t\geq 0)$ and conclude that $\mathcal{M}_\tau$ is countably generated. (See Lemma 1.3.3 in [S.&V.] for help.)

<u>Warning</u>: Unless it is otherwise stated, stopping times will be $\{\mathcal{M}_t\}$-stopping times.

Our next result plays an important role in developing criteria for determining when $M.P.((s,x);\{L_t\})$ contains at most one element (cf. Corollary (1.15) below) as well as allowing us to derive the strong Markov property as an essentially immediate consequence of such a uniqueness statement.

(1.14) <u>Theorem</u>: Let $P \in M.P.((s,x);\{L_t\})$ and a stopping time τ be given. Suppose that $\mathcal{A}$ is a sub σ-algebra of $\mathcal{M}_\tau$ with the property that $\omega \longmapsto x(\tau(\omega),\omega)$ is $\mathcal{A}$-measurable, and let $\omega \longmapsto P_\omega$ be a r.c.p.d. of $P|\mathcal{A}$. Then there is a P-null set $\Lambda \in \mathcal{A}$ such that $P_\omega \circ \theta^{-1}_{\tau(\omega)} \in M.P.(s+\tau(\omega),x(\tau(\omega),\omega);\{L_t\})$ for each $\omega \notin \Lambda$.

<u>Proof</u>: Let $\omega \longmapsto P^\tau_\omega$ be a r.c.p.d. of $P|\mathcal{M}_\tau$. Then, by Theorem (II.2.20), for each $\varphi \in C_0^\infty(\mathbb{R}^N)$:

$$(\varphi(x(t)) - \int_0^t [L_{s+\tau(\omega)+u}\varphi](x(u))du, \mathcal{M}_t, P^\tau_\omega \circ \theta^{-1}_{\tau(\omega)})$$

is a martingale for all ω outside of a P-null set $\Lambda(\varphi) \in \mathcal{M}_\tau$; and from this it is clear that there is one P-null set $\Lambda \in \mathcal{M}_\tau$ such that $P^\tau_\omega \circ \theta^{-1}_{\tau(\omega)} \in M.P.((s+\tau(\omega),x(\tau(\omega),\omega);\{L_t\})$ for all $\omega \notin \Lambda$.

To complete the proof, first note that $P_\omega(\Lambda) = 0$ for all ω outside of a P-null set $\Lambda' \in \mathcal{A}$. Second, note that $P_\omega = \int P^\tau_{\omega'} P_\omega(d\omega')$ for all ω outside of a P-null set $\Lambda'' \in \mathcal{A}$. Finally, since $P_\omega \circ \theta^{-1}_{\tau(\omega)}(x(0)=x(\tau(\omega),\omega)) = 1$ for all ω outside of a P-null set $\Lambda''' \in \mathcal{A}$, we see that $P_\omega \circ \theta^{-1}_{\tau(\omega)} \in M.P.(s+\tau(\omega),x(\tau(\omega),\omega);\{L_t\})$ for all $\omega \notin \Lambda' \cup \Lambda'' \cup \Lambda'''$. Q.E.D.

(1.15) <u>Corollary</u>: Suppose that for all $(s,x) \in [0,\infty)\times\mathbb{R}^N$, $t \geq 0$, and $P,Q \in M.P.((s,x);\{L_t\})$, $P\circ x(t)^{-1} = Q\circ x(t)^{-1}$. Then, for each $(s,x) \in [0,\infty)\times\mathbb{R}^N$, $M.P.((s,x);\{L_t\})$ contains at most one element.

Proof: Let $P,Q \in M.P.((s,x);\{L_t\})$ be given. We will prove by induction that for all $n \geq 1$ and $0 \leq t_1 < \cdots < t_n$, $P\circ(x(t_1),\ldots,x(t_n))^{-1} = Q\circ(x(t_1),\ldots,x(t_n))^{-1}$. When $n = 1$, there is nothing to prove. Next, assume that this equality holds for n and let $0 \leq t_1 < \cdots < t_{n+1}$ be given. Set $\mathcal{A} = \sigma(x(t_1),\ldots,x(t_n))$ and let $\omega \longmapsto P_\omega$ and $\omega \longmapsto Q_\omega$ be r.c.p.d.'s of $P|\mathcal{A}$ and $Q|\mathcal{A}$, respectively. Since P and Q agree on $\mathcal{A}$, there is a $\Lambda \in \mathcal{A}$ such that $P(\Lambda) = Q(\Lambda) = 0$ and both $P_\omega\circ\theta_{t_n}^{-1}$ and $Q_\omega\circ\theta_{t_n}^{-1}$ are elements of $M.P.((s+t_n,x(t_n,\omega));\{L_t\})$ for all $\omega \notin \Lambda$. In particular, $P_\omega\circ x(t_{n+1})^{-1} = Q_\omega\circ x(t_{n+1})^{-1}$ for all $\omega \notin \Lambda$; and so the inductive step is complete. Q.E.D.

(1.16) Remark: A subset $\mathcal{J}$ of bounded measurable functions on a measurable space $(E,\mathcal{B})$ is called a determining set if, for all $\mu,\nu \in \mathbf{M}_1(E)$, $\int \varphi d\mu = \int \varphi d\nu$ for all $\varphi \in \mathcal{J}$ implies that $\mu = \nu$. Now suppose that there is a determining set $\mathcal{J} \subseteq C_b(\mathbb{R}^N)$ such that for each $T > 0$ and $\varphi \in \mathcal{J}$ there is a $u = u_{t,\varphi} \in C_b^{1,2}([0,T)\times\mathbb{R}^N)$ satisfying $(\partial_t+L_t)u = 0$ in $[0,T)\times\mathbb{R}^N$ and $\lim_{t\uparrow T} u(t,\cdot) = \varphi$. Given $P \in M.P.((s,x);\{L_t\})$ and $T > 0$, we see that, for all $\varphi \in \mathcal{J}$:

$$E^P[\varphi(x(T))] = u_{s+T,\varphi}(s,x).$$

In particular, $P\circ x(T)^{-1}$ is uniquely determined for all $T > 0$ by the condition that $P \in M.P.((s,x);\{L_t\})$. Hence, for each $(s,x) \in [0,\infty)\times\mathbb{R}^N$, $M.P.((s,x);\{L_t\})$ contains at most one element. Similarly, suppose that for each $\varphi \in \mathcal{J}$ there is a $u = u_{t,\varphi} \in C_b^{1,2}([0,T)\times\mathbb{R}^N)$ satisfying $(\partial_t+L_t)u = \varphi$ in $[0,T)\times\mathbb{R}^N$

and $\lim_{t\uparrow T} u(t,\cdot) = 0$. Then $P \in M.P.((s,x);\{L_t\})$ implies that:

$$E^P\left[\int_0^T \varphi(x(t))dt\right] = u_{s+T,\varphi}(s,x)$$

for all $T > 0$ and $\varphi \in \mathcal{J}$. From this it is clear that if $P,Q \in M.P.((s,x);\{L_t\})$, then $P\circ x(t)^{-1} = Q\circ x(t)^{-1}$ for all $t > 0$, and so $M.P.((s,x);\{L_t\})$ contains at most one element. It should be clear that this last remark is simply a re-statement of the result in Corollary (II.1.13).

We now introduce a construction which will serve us well when it comes to proving the existence of solutions to martingale problems as well as reducing the question of uniqueness to local considerations.

(1.17)<u>Lemma</u>: Let $T > 0$ and $\psi \in C([0,T];\mathbb{R}^N)$ be given. Suppose that $Q \in \mathbf{M}_1(\Omega)$ satisfies $Q(x(0)=\psi(T)) = 1$. Then there is a unique $R = \delta_\psi \otimes_T Q \in \mathbf{M}_1(\Omega)$ such that $R(A\cap\theta_T^{-1}B) = \chi_A(\psi\restriction[0,T])Q(B)$ for all $A \in \mathcal{M}_T$ and $B \in \mathcal{M}$.

<u>Proof</u>: The uniqueness assertion is clear. To prove the existence, set $\tilde{R} = \delta_\psi \times Q$ on $\Omega\times\Omega$ and define $\Phi:\Omega\times\Omega\longrightarrow\Omega$ so that

$$x(t,\Phi(\omega,\omega')) = \begin{array}{l} x(t,\omega) \text{ if } t \in [0,T] \\ x(t-T,\omega') - x(T,\omega') + x(T,\omega) \text{ if } t > T. \end{array}$$

Then $R = \tilde{R}\circ\Phi^{-1}$ has the required property. Q.E.D.

(1.18)<u>Theorem</u>: Let τ be a stopping time and suppose that $\omega \in \Omega \longmapsto Q_\omega \in \mathbf{M}_1(\Omega)$ is an $\mathcal{M}_\tau$-measurable map satisfying $Q_\omega(x(\tau(\omega))=x(\tau(\omega),\omega)) = 1$ for each $\omega \in \Omega$. Given $P \in \mathbf{M}_1(\Omega)$, there is a unique $R = P\otimes_\tau Q_\cdot \in \mathbf{M}_1(\Omega)$ such that $R\restriction\mathcal{M}_\tau = P\restriction\mathcal{M}_\tau$ and $\omega\longmapsto\delta \otimes \quad Q$ is a r.c.p.d. of $R|\mathcal{M}$. In addition, suppose that $(\omega,t,\omega') \in \Omega\times[0,\infty)\times\Omega \longmapsto Y_\omega(t,\omega') \in \mathbb{R}^1$ is a map such that, for each $T > 0$, $(\omega,t,\omega') \in \Omega\times[0,T]\times\Omega \longmapsto Y_\omega(t,\omega')$ is $\mathcal{M}_\tau\times\mathcal{B}_{[0,T]}\times\mathcal{M}_T$-measurable, and, for each $\omega,\omega' \in \Omega$, $t \longmapsto Y_\omega(t,\omega)$ is a right continuous function with $Y_\omega(0,\omega') = 0$. Given a right continuous progressively measurable function X: $[0,\infty)\times\Omega \longrightarrow \mathbb{R}^1$, define $Z = X\otimes_\tau Y_\cdot$ by:

$$Z(t,\omega) = \begin{array}{l} X(t,\omega) \text{ if } t \in [0,\tau(\omega)) \\ X(\tau(\omega),\omega) + Y_\omega(t-\tau(\omega),\theta_{\tau(\omega)}\omega) \text{ if } t > \tau(\omega). \end{array}$$

Then, Z is a right continuous progressively measurable function. Moreover, if $Z(t) \in L^1(R)$ for all $t \geq 0$, then $(Z(t),\mathcal{M}_t,R)$ is a martingale if and only if $(X(t\wedge\tau),\mathcal{M}_t,P)$ is a martingale and $(Y_\omega(t),\mathcal{M}_t,Q_\omega)$ is a martingale for P-almost every $\omega \in \Omega$.

<u>Proof</u>: The uniqueness of R is obvious; to prove existence, set $R = \int\delta_\omega\otimes_{\tau(\omega)}Q_\omega P(d\omega)$ and note that

$$R(A\cap B) = \int_A \delta_\omega\otimes_{\tau(\omega)}Q_\omega(B)P(d\omega)$$

for all $A \in \mathcal{M}_\tau$ and $B \in \mathcal{M}$. To see that Z is progressively measurable, define

$$\tilde{Z}(t,\omega,\omega') = X(t,\omega) + Y_\omega((t-\tau(\omega))\vee 0,\theta_{\tau(\omega)}\omega')$$

and check that $\tilde{Z}$ is $\{\mathcal{M}_t\times\mathcal{M}_t\}$-progressively measurable and that $Z(t,\omega) = \tilde{Z}(t,\omega,\omega)$. To complete the proof, simply apply Theorem (II.2.20). Q.E.D.

We now turn to the problem of constructing solutions. Let a: $[0,\infty)\times\mathbb{R}^N \longrightarrow S^+(\mathbb{R}^N)$ and b: $[0,\infty)\times\mathbb{R}^N \longrightarrow \mathbb{R}^N$ be bounded continuous functions and define $\{L_t\}$ accordingly. For $(s,x) \in [0,\infty)\times\mathbb{R}^N$, define $\Psi^{a,b}_{s,x}$: $\Omega \longrightarrow \Omega$ so that:

$$x(t,\Psi^{a,b}_{s,x}(\omega)) = x + a(s,x)^{1/2}x(t,\omega) + b(s,x)t,\ t \geq 0,$$

and define $\mathcal{W}^{a,b}_{s,x} = \mathcal{W}\circ(\Psi^{a,b}_{s,x})^{-1}$. It is then easy to check that $\bar{x} \in (Mart^2_c(\{\mathcal{M}_t\},\mathcal{W}^{a,b}_{s,x}))^N$, where $\bar{x}(t) \equiv x(t) - b(s,x)t$, and that $\langle\langle\bar{x},\bar{x}\rangle\rangle(t) = a(s,x)t,\ t \geq 0$. Next, for $n \geq 1$, define $P^{n,k}_x$ for $k \geq 0$ so that $P^{n,0}_x = \mathcal{W}^{a,b}_{0,x}$ and

$$P^{n,k}_x = P^{n,k-1}_x \otimes_{\frac{k-1}{n}} \mathcal{W}^{a,b}_{\frac{k-1}{n},x(\frac{k-1}{n})}$$

for $k \geq 1$. Set $P^n_x = P^{n,n^2}_x$ and define:

$$L^n_t = 1/2\sum_{i,j=1}^{N} a^{i,j}\Big(\frac{[nt]\wedge n^2}{n},x\Big(\frac{[nt]\wedge n^2}{n}\Big)\Big)\partial_{x^i}\partial_{x^j} + \sum_{i=1}^{N} b^i\Big(\frac{[nt]\wedge n^2}{n},x\Big(\frac{[nt]\wedge n^2}{n}\Big)\Big)\partial_{x^i}.$$

Then $P^n_x \in M.P.((0,x);\{L^n_t\})$ (cf. exercise (1.9) above). In particular, for each $T > 0$ there is a $C(T) < \infty$ such that:

$$\sup_{n\geq 1}\sup_{x} E^{P^n_x}\Big[|x(t)-x(s)|^4\Big] \leq C(T)(v-u)^2,\ 0 \leq s,t \leq T.$$

Hence, $\{P^n_x\colon n \geq 1\}$ is relatively compact in $\mathbf{M}_1(\Omega)$ for each $x \in \mathbb{R}^N$. Because, for each $\varphi \in C^\infty_0(\mathbb{R}^N)$,

$$[L^n_t\varphi](x(t,\omega)) \longrightarrow [L_t\varphi](x(t,\omega))$$

uniformly for (t,ω) in compact subsets of $[0,\infty)\times\Omega$, our construction will be complete once we have available the result contained in the following exercise.

(1.19) <u>Exercise</u>: Let E be a Polish space and suppose that $\mathcal{F} \subseteq C_b(E)$ be a uniformly bounded family of functions which are equi-continuous on each compact subset of E. Show that if $\mu_n \longrightarrow \mu$ in $\mathbf{M}_1(\Omega)$, then $\int \varphi d\mu_n \longrightarrow \int \varphi d\mu$ uniformly for $\varphi \in \mathcal{F}$. In particular, if $\{\varphi_n\} \subseteq C_b(E)$ is uniformly bounded and $\varphi_n \longrightarrow \varphi$ uniformly on compacts, then $\int \varphi_n d\mu_n \longrightarrow \int \varphi d\mu$ whenever $\mu_n \longrightarrow \mu$ in $\mathbf{M}_1(\Omega)$.

Referring to the paragraph preceding exercise (1.19), we now see that if $\{P_x^{n'}\}$ is any convergent subsequence of $\{P_x^n\}$ and if P denotes its limit, then $P(x(0)=x) = 1$ and, for all $0 \leq t_1 < t_2$, all $\mathcal{M}_{t_1}$-measurable $\Phi \in C_b(\Omega)$, and all $\varphi \in C_0^\infty(\mathbb{R}^N)$:

$$E^P\left[\left[\varphi(x(t_2))-\varphi(x(t_1))\right]\Phi\right] = E^P\left[\left[\int_{t_1}^{t_2}[L_t\varphi](x(t))dt\right]\Phi\right].$$

From this it is clear that $P \in M.P.((0,x);\{L_t\})$. By replacing a and b with $a(s+\cdot,*)$ and $b(s+\cdot,*)$, we also see that there is at least one $P \in M.P.((s,x);\{L_t\})$ for each $(s,x) \in [0,\infty)\times\mathbb{R}^N$. In other words, we have now proved the following existence theorem.

(1.20) <u>Theorem</u>: Let $a: [0,\infty)\times\mathbb{R}^N \longrightarrow S^+(\mathbb{R}^N)$ and $b: [0,\infty)\times\mathbb{R}^N \longrightarrow \mathbb{R}^N$ be bounded continuous functions and define $\{L_t\}$ accordingly. Then, for each $(s,x) \in [0,\infty)\times\mathbb{R}^N$ there is at least one element of $M.P.((s,x);\{L_t\})$.

(1.21) <u>Exercise</u>: Suppose that a and b are bounded, measurable and have the property that $x \in \mathbb{R}^N \longmapsto \int_0^T a(t,x)dt$ and $x \in \mathbb{R}^N \longmapsto \int_0^T b(t,x)dt$ are continuous for each $T > 0$. Show that the corresponding martingale problem still has a solution starting at each $(s,x) \in [0,\infty)\times\mathbb{R}^N$.

We now have the basic existence result and uniqueness criteria for martingale problems coming from diffusion operators (i.e. operators of the sort in (II.1.13)). However, before moving on, it may be useful to record a summary of the rewards which follow from proving that a martingale problem is <u>well-posed</u> in the sense that precisely one solution exists for each starting point (s,x).

(1.22) <u>Theorem</u>: Let a and b be bounded measurable functions, suppose that the martingale problem for the corresponding $\{L_t\}$ is well posed, and let $\{P_{s,x}: (s,x) \in \mathbb{R}^N\}$ be the associated family of solutions. Then $(s,x)\longmapsto P_{s,x}$ is measurable; and, for all stopping times τ, $P_{s,x} = P_{s,x}\otimes_\tau P_{\tau,x(\tau)}$. In particular, $\omega\longmapsto\delta_\omega\otimes_{\tau(\omega)}P_{\tau(\omega),x(\tau(\omega),\omega)}$ is a r.c.p.d. of $P_{s,x}|\mathcal{M}_\tau$ for each stopping times τ. Finally, if $x \in \mathbb{R}^N\longmapsto\int_0^T a(t,x)dt$ and $x \in \mathbb{R}^N\longmapsto\int_0^T b(t,x)dt$ are continuous for each $T > 0$, then $(s,x) \in [0,\infty)\times\mathbb{R}^N\longmapsto P_{s,x} \in \mathbf{M}_1(\Omega)$ is continuous.

<u>Proof</u>: The measurability of $(s,x)\longmapsto P_{s,x}$ is proved in Theorem (1.12). Next, let τ be a stopping time. Then, by Theorem (1.18), it is easy to check that, for all $\varphi \in C_0^\infty(\mathbb{R}^N)$, $(X_{s,\varphi}(t),\mathcal{M}_t,P_{s,x}\otimes_\tau P_{\tau,x(\tau)})$ is a martingale (cf. (1.1) for the notation $X_{s,\varphi}$). Hence $P_{s,x}\otimes_\tau P_{\tau,x(\tau)} \in$ M.P.$((s,x);\{L_t\})$; and so, by uniqueness, $P_{s,x}\otimes_\tau P_{\tau,x(\tau)} = P_{s,x}$.

Finally, suppose that $x \in \mathbb{R}^N \longmapsto \int_0^T a(t,x)dt$ and $x \in \mathbb{R}^N \longmapsto \int_0^T b(t,x)dt$ are continuous for each $T > 0$. Then, for each $\varphi \in C_0^\infty(\mathbb{R}^N)$, $(s,t,\omega) \in [0,\infty)\times[0,\infty)\times\Omega \longmapsto X_{s,\varphi}(t,\omega)$ is continuous. Now let $(s_n,x_n) \longrightarrow (s,x)$ and assume that $P_{s_n,x_n} \longrightarrow P$. Then, by exercise (1.19):

$$\int X_{s,\varphi}(t)\Phi dP = \lim_{n\to\infty} \int X_{s_n,\varphi}(t)\Phi dP_{s_n,x_n}$$

for all $t \geq 0$, $\varphi \in C_0^\infty(\mathbb{R}^N)$, and $\Phi \in C_b(\Omega)$. Hence, $P \in$ M.P.$((s,x);\{L_t\})$, and so $P = \lim_{n\to\infty} P_{s_n,x_n}$. At the same time, by (1.6) and Kolmogorov's criterion, $\{P_{s,x}: s\geq 0 \text{ and } |x|\leq R\}$ is relatively compact in $\mathbf{M}_1(\Omega)$ for each $R \geq 0$; and combined with the preceding, this leads immediately to the conclusion that $(s,x) \longmapsto P_{s,x}$ is continuous. Q.E.D.

(1.23)<u>Exercise</u>: For each $n \geq 1$, let a_n and b_n be given bounded measurable coefficients and let P^n be a solution to the corresponding martingale problem starting from some point (s_n,x_n). Assume that $(s_n,x_n) \longrightarrow (s,x)$ and that $a_n \longrightarrow a$ and $b_n \longrightarrow b$ uniformly on compacts, where a and b are bounded measurable coefficients such that $x \longmapsto \int_0^T a(t,x)dt$ and $x \longmapsto \int_0^T b(t,x)dt$ are continuous. If the martingale problem corresponding to a and b starting from (s,x) has precisely one solution $P_{s,x}$, show that $P^n \longrightarrow P_{s,x}$.

2. The Martingale Problem and Stochastic Integral Equations:

Let $\mathbf{a}$: $[0,\infty)\times\mathbb{R}^N \longrightarrow S^+(\mathbb{R}^N)$ and b: $[0,\infty)\times\mathbb{R}^N \longrightarrow \mathbb{R}^N$ be bounded measurable functions and define $t \longmapsto L_t$ accordingly. When $\mathbf{a} \equiv 0$, we saw (cf. the remark following (1.2)) that $P \in M.P.((s,x);\{L_t\})$ if and only if $x(T) = x + \int_0^T b(s+t,x(t))dt$, $T \geq 0$, (a.s.,P). We now want to see what can be said when $\mathbf{a}$ does not vanish identically. In order to understand what we have in mind, assume that $N = 1$ and that $\mathbf{a}$ never vanishes. Given $P \in M.P.((s,x);\{L_t\})$, define

$$\beta(T) = \int_0^T \mathbf{a}^{-1/2}(s+t,x(t))d\bar{x}(t),\ T \geq 0,$$

where $\bar{x}(T) = x(T) - \int_0^T b(s+t,x(t))dt$, $T \geq 0$. Then, $\langle\beta,\beta\rangle(dt) = (\mathbf{a}^{-1/2}(s+t,x(t)))^2\mathbf{a}(s+t,x(t))dt = dt$, and so $(\beta(t),\mathcal{M}_t,P)$ is a 1-dimensional Brownian motion. In addition:

$$\bar{x}(T) - x = \int_0^T d\bar{x}(t) = \int_0^T \mathbf{a}^{1/2}(s+t,x(t))d\beta(t),\ T \geq 0,\ (a.s.,P);$$

and so $x(\cdot)$ satisfies the <u>stochastic integral equation</u>:

$$x(T) = x + \int_0^T \mathbf{a}^{1/2}(s+t,x(t))d\beta(t) + \int_0^T b(s+t,x(t))dt,\ T \geq 0,$$

(a.s.,P). Our first goal in this section is to generalize the preceding representation theorem. However, before doing so, we must make a brief digression into the theory of stochastic integration with respect to vector-valued martingales.

Referring to the notation introduced in section 2 of Chapter II, let $d \in Z^+$ and $X \in (Mart_c^2(\{\mathcal{F}_t\},P))^d$ be given.

Define $L^2_{loc}(\{\mathcal{F}_t\},\langle\langle X,X\rangle\rangle,P)$ to be the space of $\{\mathcal{F}_t\}$-progressively measurable θ: $[0,\infty)\times E\longrightarrow\mathbb{R}^d$ such that:

$$E^P\left[\int_0^T(\theta(t),\langle\langle X,X\rangle\rangle(dt)\theta(t))_{\mathbb{R}^d}\right] < \infty,\ T \geq 0.$$

Note that $(L^2_{loc}(\{\mathcal{F}_t\},\mathrm{Trace}\langle\langle X,X\rangle\rangle,P))^d$ can be identified as a dense subspace of $L^2_{loc}(\{\mathcal{F}_t\},\langle\langle X,X\rangle\rangle,P)$ (to see the density, simply take $\theta_n(t) \equiv \chi_{[0,n)}(|\theta(t)|)\theta(t)$ to approximate θ in $L^2_{loc}(\{\mathcal{F}_t\},\langle\langle X,X\rangle\rangle,P)$). Next, for $\theta \in (L^2_{loc}(\{\mathcal{F}_t\},\mathrm{Trace}\langle\langle X,X\rangle\rangle,P))^d$, define:

$$\int_0^T\theta(t)dX(t) = \sum_{i=1}^d\int_0^T\theta_i(t)dX^i(t),\ T \geq 0; \tag{2.1}$$

and observe that:

$$E^P\left[\sup_{0\leq t\leq T}\left|\int_0^t\theta dX\right|^2\right] \leq 4E^P\left[\left|\int_0^T\theta dX\right|^2\right]$$

$$= 4E^P\left[\int_0^T(\theta(t),\langle\langle X,X\rangle\rangle(dt)\theta(t))_{R^d}\right].$$

Hence there is a unique continuous mapping

$$\theta \in L^2_{loc}(\{\mathcal{F}_t\},\langle\langle X,X\rangle\rangle,P)\longmapsto\int_0^\cdot\theta dX \in \mathrm{Mart}^2_c(\{\mathcal{F}_t\},P)$$

such that $\int_0^\cdot\theta dX$ is given by (2.1) whenever $\theta \in (L^2_{loc}(\{\mathcal{F}_t\},\mathrm{Trace}\langle\langle X,X\rangle\rangle,P))^d$.

(2.2)Exercise: Given $\theta \in L^2_{loc}(\{\mathcal{F}_t\},\langle\langle X,X\rangle\rangle,P)$, show that $\int_0^\cdot\theta dX$ is the unique $Y \in \mathrm{Mart}^2_c(\{\mathcal{F}_t\},P)$ such that:

$$\langle Y,\int_0^\cdot\eta dX\rangle(dt) = (\theta(t),\langle\langle X,X\rangle\rangle(dt)\eta(t))_{\mathbb{R}^d},\ (\text{a.s.},P),$$

for all $\eta \in L^2_{loc}(\{\mathcal{F}_t\},\langle\langle X,X\rangle\rangle,P)$. In particular, conclude that if $\theta \in L^2_{loc}(\{\mathcal{F}_t\},\langle\langle X,X\rangle\rangle,P))$ and τ is an $\{\mathcal{F}_t\}$-stopping time, then:

$$\int_0^{T\wedge\tau}\theta dX = \int_0^T \chi_{[0,\tau)}(t)\theta(t)dX(T),\ T \geq 0,\ (\text{a.s.},P).$$

Next, suppose that σ: $[0,\infty)\times E \longrightarrow \text{Hom}(R^d;R^N)$ is an $\{\mathcal{F}_t\}$-progressively measurable map which satisfies:

$$E^P\left[\int_0^T \text{Trace}(\sigma(t)\langle\langle X,X\rangle\rangle(dt)\sigma(t)^\dagger)\right] < \infty,\ T \geq 0. \tag{2.3}$$

We then define $\underline{\int_0^\cdot \sigma(t)dX(t)}$ $\in (\text{Mart}_c^2(\{\mathcal{F}_t\},P))^N$ so that:

$$(\theta,\int_0^\cdot \sigma dX)_{\mathbb{R}^N} = \int_0^\cdot (\sigma^\dagger\theta)dX,\ (\text{a.s.},P),$$

for each $\theta \in \mathbb{R}^N$.

(2.4)<u>Exercise</u>: Let $X \in (\text{Mart}_c^2(\{\mathcal{F}_t\},P))^d$ and an $\{\mathcal{F}_t\}$-progressively measurable σ: $[0,\infty)\times E \longrightarrow \text{Hom}(\mathbb{R}^d;\mathbb{R}^N)$ satisfying (2.3) be given.

i) If $Y \in (\text{Mart}_c^2(\{\mathcal{F}_t\},P))^e$ and τ: $[0,\infty)\times E \longrightarrow \text{Hom}(\mathbb{R}^e;\mathbb{R}^N)$ is an $\{\mathcal{F}_t\}$-progressively measurable function which satisfies:

$$E^P\left[\int_0^T \text{Trace}(\tau(t)\langle\langle Y,Y\rangle\rangle(dt)\tau(t)^\dagger) < \infty,\ T > 0,\right.$$

show that

$$\int_0^\cdot [\sigma,\tau]d\begin{bmatrix}X\\Y\end{bmatrix} = \int_0^\cdot \sigma dX + \int_0^\cdot \tau dY,\ (\text{a.s.},P)$$

and that

$$\langle\langle\int_0^\cdot \sigma dX,\int_0^\cdot \tau dY\rangle\rangle(dt) = \sigma(t)\langle\langle X,Y\rangle\rangle(dt)\tau(t)^\dagger,\ (\text{a.s.},P)$$

where $\langle\langle X,Y\rangle\rangle \equiv ((\langle X^i,Y^j\rangle))_{\substack{1\leq i\leq d\\ 1\leq j\leq e}}$.

ii) Show that $\int_0^\cdot \sigma dX$ is the unique $Y \in (\text{Mart}_c^2(\{\mathcal{F}_t\},P))^N$ such that $Y(0) = 0$ and

$$\langle\langle \begin{bmatrix} X \\ Y \end{bmatrix}, \begin{bmatrix} X \\ Y \end{bmatrix} \rangle\rangle(dt) = \begin{bmatrix} I \\ \sigma(t) \end{bmatrix} \langle\langle X, X \rangle\rangle(dt)[I, \sigma(t)^{\dagger}],$$

(a.s.,P).

iii) Next, show that if τ: $[0,\infty)\times E \longrightarrow \mathrm{Hom}(\mathbb{R}^N;\mathbb{R}^M)$ is an $\{\mathcal{F}_t\}$-progressively measurable function such that (2.3) is satisfied both by σ and by $\tau\sigma$, then:

$$\int_0^{\cdot} \tau(t)d\left[\int_0^t \sigma dX\right] = \int_0^{\cdot} \tau\sigma dX, \quad (a.s.,P).$$

The following lemma addresses a rather pedantic measurability question.

(2.5)<u>Lemma</u>: Let $\mathbf{a} \in S^+(\mathbb{R}^N)$, denote by π the orthognal projection of $\mathbb{R}^N$ onto Range(**a**), and let $\tilde{\mathbf{a}}$ be the element of $S^+(\mathbb{R}^N)$ satisfying $\tilde{\mathbf{a}}\mathbf{a} = \mathbf{a}\tilde{\mathbf{a}} = \pi$. Then $\pi = \lim_{\epsilon\downarrow 0} (\mathbf{a} + \epsilon I)^{-1}\mathbf{a}$ and $\tilde{\mathbf{a}} = \lim_{\epsilon\downarrow 0} (\mathbf{a} + \epsilon I)^{-1}\pi$. Next, suppose that $\sigma \in \mathrm{Hom}(\mathbb{R}^d;\mathbb{R}^N)$ and that $\mathbf{a} = \sigma\sigma^{\dagger}$. Let π_σ denote the orthogonal projection of $\mathbb{R}^N$ onto Range($\sigma^{\dagger}$). Then Range(**a**) = Range(σ) and $\sigma^{\dagger}\tilde{\mathbf{a}}\sigma = \pi_\sigma$. In particular, $\mathbf{a}\longmapsto\pi$ and $\mathbf{a}\longmapsto\tilde{\mathbf{a}}$ are measurable functions of **a**, and $\sigma\longmapsto\pi_\sigma$ is a measurable function of σ.

<u>Proof</u>: Set $\mathbf{a}_\epsilon = (\mathbf{a} + \epsilon I)^{-1}$ and $\pi_\epsilon = \mathbf{a}_\epsilon\mathbf{a}$. Then $0 \le \pi_\epsilon \le I$. Moreover, if $\eta \in \mathrm{Range}(\mathbf{a})^{\perp}$, then $\eta \in$ Null(**a**) and so $\pi_\epsilon\eta = 0$; whereas, if $\eta \in$ Range(**a**), then there is a $\eta = \mathbf{a}\xi$, and so $\pi_\epsilon\eta = \eta - \epsilon\pi_\epsilon\xi \longrightarrow \eta$ as $\epsilon\downarrow 0$. Hence, $\pi_\epsilon \longrightarrow \pi$ as $\epsilon\downarrow 0$. Also, if $\eta \in \mathrm{Range}(\mathbf{a})^{\perp}$, then $\mathbf{a}_\epsilon\pi\eta = 0$; and if $\eta \in$ Range(**a**), then there is a $\xi \in$ Range(**a**) such that $\eta = \mathbf{a}\xi$, and so $\mathbf{a}_\epsilon\pi\eta = \mathbf{a}_\epsilon\eta = \pi_\epsilon\xi \longrightarrow \xi$ as $\epsilon\downarrow 0$. Hence, $\mathbf{a}_\epsilon\pi \longrightarrow \tilde{\mathbf{a}}$ as $\epsilon\downarrow 0$. Since, for each $\epsilon > 0$, $\mathbf{a}\longmapsto\mathbf{a}_\epsilon$ is a smooth map, we now see that $\mathbf{a}\longmapsto\pi$ and $\mathbf{a}\longmapsto\tilde{\mathbf{a}}$ are measurable maps. Now suppose that $\mathbf{a} = \sigma\sigma^{\dagger}$. Clearly

$\text{Range}(a) \subseteq \text{Range}(\sigma)$. On the other hand, if $\eta \in \text{Range}(\sigma)$, then there exists a $\xi \in \text{Null}(\sigma)^{\perp} = \text{Range}(\sigma^{\dagger})$ such that $\eta = \sigma\xi$. Hence, choosing η' so that $\xi = \sigma^{\dagger}\eta'$, we then have $\eta = \sigma\sigma^{\dagger}\xi = a\eta'$; from which we conclude that $\text{Range}(a) = \text{Range}(\sigma)$. Finally, to see that $\sigma^{\dagger}\tilde{a}\sigma = \pi_{\sigma}$, note that if $\eta \in \text{Range}(\sigma^{\dagger})^{\perp}$, then $\eta \in \text{Null}(\sigma)$ and so $\sigma^{\dagger}\tilde{a}\sigma\eta = 0$. On the other hand, if $\eta \in \text{Range}(\sigma^{\dagger})$, then there is a $\xi \in \text{Null}(\sigma^{\dagger})^{\perp} = \text{Range}(\sigma) = \text{Range}(a)$ such that $\eta = \sigma^{\dagger}\xi$, and so $\sigma^{\dagger}\tilde{a}\sigma\eta = \sigma^{\dagger}\tilde{a}a\xi = \sigma^{\dagger}\pi\xi = \sigma^{\dagger}\xi = \eta$. Hence, $\sigma^{\dagger}\tilde{a}\sigma = \pi_{\sigma}$. Q.E.D.

We are at last ready to prove the representation theorem alluded to above.

(2.6)<u>Theorem</u>: Let $a: [0,\infty)\times\mathbb{R}^N \longrightarrow S^+(\mathbb{R}^N)$ and $b: [0,\infty)\times\mathbb{R}^N \longrightarrow \mathbb{R}^N$ be bounded measurable functions and suppose that $P \in M.P.((s,x);\{L_t\})$ for some $(s,x) \in [0,\infty)\times\mathbb{R}^N$. Given a measurable $\sigma: [0,\infty)\times\mathbb{R}^N \longrightarrow \text{Hom}(\mathbb{R}^d;\mathbb{R}^N)$ satisfying $a = \sigma\sigma^{\dagger}$, there is a d-dimensional Brownian motion $(\beta(t),\mathcal{F}_t,Q)$ on some probability space $(E,\mathcal{F},Q)$ and a continuous $\{\mathcal{F}_t\}$-progressively measurable function $X: [0,\infty)\times E \longrightarrow \mathbb{R}^N$ such that

$$X(T) = x + \int_0^T \sigma(s+t,X(t))d\beta(t) + \int_0^T b(s+t,X(t))dt, \qquad (2.7)$$

$T \geq 0$, (a.s.,Q), and $P = Q\circ X(\cdot)^{-1}$. In particular, if $d = N$ and a is never singular (i.e. $a(t,y) > 0$ for all $(t,y) \in [0,\infty)\times\mathbb{R}^N$), then we can take $E = \Omega$, $Q = P$, $X(\cdot) = x(\cdot)$ and $\beta(T) = \int_0^T \sigma^{\dagger}a^{-1}(s+t,x(t))d\bar{x}(t)$, $T \geq 0$, where

$$\bar{x}(T) \equiv x(T) - \int_0^T b(s+t,x(t))dt, \quad T \geq 0.$$

<u>Proof</u>: Let $E = C([0,\infty);\mathbb{R}^N\times\mathbb{R}^d) = C([0,\infty);\mathbb{R}^N)\times C([0,\infty);\mathbb{R}^d)$, $\mathcal{F} = \mathcal{B}_E$, and $Q = P\times\mathcal{W}$, where $\mathcal{W}$ denotes d-dimensional Wiener measure on $\Omega_d \equiv C([0,\infty);\mathbb{R}^d)$. Given $\xi \in E$, let $Z(T,\xi) = \begin{bmatrix}X(T,\xi)\\Y(T,\xi)\end{bmatrix} \in \mathbb{R}^N\times\mathbb{R}^d$ denote the position of the path ξ at time T, set $\mathcal{F}_t = \sigma(Z(u)\colon 0\le u\le t)$ for $t \ge 0$, and note that (by the second part of exercise (II.2.31)) $\bar{Z} \in (\mathrm{Mart}_c^2(\{\mathcal{F}_t\},P))^{N+d}$ with

$$\langle\langle\bar{Z},\bar{Z}\rangle\rangle(dt) = \begin{bmatrix}a(s+t,X(t)) & 0\\ 0 & I_{\mathbb{R}^d}\end{bmatrix}dt,$$

where $\bar{Z}(T) \equiv \begin{bmatrix}\bar{X}(T)\\Y(T)\end{bmatrix}$ and $\bar{X}(T) \equiv X(T) - \int_0^T b(s+t,X(t))dt$. Next, define $\pi(t,y)$ and $\pi_\sigma(t,y)$ to be the orthogonal projections of $\mathbb{R}^N$ and $\mathbb{R}^d$ onto Range(a(t,y)) and Range($\sigma^\dagger(t,y)$), respectively. Set $\pi_\sigma^\perp = I_{\mathbb{R}^d} - \pi_\sigma$, and define $\beta(\cdot)$ by

$$\beta(T) = \int_0^T[\sigma^\dagger\tilde{a},\pi_\sigma^\perp](s+t,X(t))d\bar{Z}(t),\ T \ge 0.$$

Then:

$$\langle\langle\beta,\beta\rangle\rangle(dt) = [\sigma^\dagger\tilde{a},\pi_\sigma^\perp]\begin{bmatrix}a(s+t,X(t)) & 0\\ 0 & I_{\mathbb{R}^d}\end{bmatrix}\begin{bmatrix}\tilde{a}\sigma\\ \pi_\sigma^\perp\end{bmatrix}dt$$

$$= (\sigma^\dagger\tilde{a}a\tilde{a}\sigma + \pi_\sigma^\perp)(s+t,X(t)) = I_{\mathbb{R}^d}dt,$$

since $\sigma^\dagger\tilde{a}a\tilde{a}\sigma = \sigma^\dagger\pi\tilde{a}\sigma = \sigma^\dagger\tilde{a}\sigma = \pi_\sigma$. Hence, $(\beta(t),\mathcal{F}_t,Q)$ is a d-dimensional Brownian motion. Moreover, since $\sigma\sigma^\dagger\tilde{a} = a\tilde{a} = \pi$, we see that:

$$\int_0^T\sigma(s+t,X(t))d\beta(t) = \int_0^T\pi(s+t,X(t))d\bar{X}(t)$$

$$= X(T) - x - \int_0^T b(s+t,X(t))dt - \int_0^T\pi^\perp(s+t,X(t))d\bar{X}(t),$$

where $\pi^\perp \equiv I_{\mathbb{R}^N} - \pi$. At the same time:

$$<<\int_0^{\cdot}\pi^{\perp}d\overline{X},\int_0^{\cdot}\pi^{\perp}d\overline{X}>>(dt) = \pi^{\perp}a\pi^{\perp}(s+t,X(t))dt = 0,$$

and so $\int_0^{\cdot}\pi^{\perp}d\overline{X} = 0$ (a.s.,Q). We have therefore proved that $X(\cdot)$ satisfies (2.7) with this choice of $(\beta(t),\mathcal{F}_t,Q)$; and clearly $P = Q\circ X(\cdot)^{-1}$. Moreover, if $N = d$ and $\mathbf{a}$ is never singular, then $\pi_{\sigma}^{\perp} \equiv 0$, and so we could have carried out the whole procedure on $(\Omega,\mathcal{M},P)$ instead of $(E,\mathcal{F},Q)$. Q.E.D.

(2.8) Remark: It is not true in general that the $X(\cdot)$ in (2.7) is a measurable function of the $\beta(\cdot)$ in that equation. To dramatize this point, we look at the case when $N = d = 1$, $\mathbf{a} \equiv 1$, $b \equiv 0$, $s = 0$, $x = 0$, and $\sigma(x) = \text{sgn}(x)$. Obviously, $(x(t),\mathcal{M}_t,P)$ is a 1-dimensional Brownian motion; and, by i) in exercise (II.3.22), we see that in this case: $\beta(T) = |x(T)| - 2\ell(T,0)$ (a.s.,P), where $\ell(\cdot,0)$ is the local time of $x(\cdot)$ at 0. In particular, since $\ell(T,0) = \lim_{\epsilon\downarrow 0}\int_0^T \chi_{[0,\epsilon)}(|x(t)|)dt$, (a.s.,P), $\beta(\cdot)$ is measurable with respect to the P-completion $\mathcal{A}$ of $\sigma(|x(t)|: t\geq 0)$. On the other hand, if $x(\cdot)$ were $\mathcal{A}$-measurable, then, there would exist a measurable function $\Phi: C([0,\infty);[0,\infty))\longrightarrow\mathbb{R}^1$ such that $x(1,\omega) = \Phi(|x(\cdot,\omega)|)$ for every $\omega \in \Omega$ which is not in a P-null set Λ. Moreover, since $P(-\Lambda) = P(\Lambda)$, we could assume that $\Lambda = -\Lambda$. But this would mean that $x(1,\omega) = \Phi(|x(\cdot,\omega)|) = \Phi(|x(\cdot,-\omega)|) = x(1,-\omega) = -x(1,\omega)$ for $\omega \notin \Lambda$; and so we would have that $P(x(1)=0) = 1$, which is clearly false. Hence, $x(\cdot)$ is not $\mathcal{A}$-measurable.

In spite of the preceding remark, equation (2.7) is

often a very important tool with which to study solutions to martingale problems. Its usefulness depends on our knowing enough about the smoothness of σ and b in order to conclude from (2.7) that $X(\cdot)$ can be expressed as a measurable functional of $\beta(\cdot)$. The basic result in this direction is contained in the next statement.

(2.9) <u>Theorem</u> (Itô): Let σ: $[0,\infty)\times\mathbb{R}^N \longrightarrow \mathrm{Hom}(\mathbb{R}^N;\mathbb{R}^d)$ and b $: [0,\infty)\times\mathbb{R}^N \longrightarrow \mathbb{R}^N$ are measurable functions with the property that, for each $T > 0$, there exists a $C(T) < \infty$ such that:

$$\begin{aligned} &\sup_{0\le t\le T}\|\sigma(t,0)\|_{H.S.} \vee |b(t,0)| \le C(T) \\ &\sup_{0\le t\le T}\|\sigma(t,y') - \sigma(t,y)\|_{H.S.} \vee |b(t,y') - b(t,y)| \\ &\qquad\qquad \le C(T)|y' - y|,\ y,y' \in \mathbb{R}^N. \end{aligned} \tag{2.10}$$

(Here, and throughout, $\|\cdot\|_{H.S.}$ denotes the Hilbert-Schmidt norm.) Denote by $\mathcal{W}$ the standard d-dimensional Wiener measure on $(\Omega,\mathcal{M})$. Then there is for each $(s,x) \in [0,\infty)\times\mathbb{R}^N$ a right continuous, $\{\mathcal{M}_t\}$-progressively measurable map $\Phi_{s,x}$: $[0,\infty)\times\Omega \longrightarrow \mathbb{R}^N$ such that

$$\Phi_{s,x}(T) = x + \int_0^T \sigma(s+t,\Phi_{s,x}(t))dx(t) + \int_0^T b(s+t,\Phi(t))dt,\ T \ge 0,$$

(a.s.,$\mathcal{W}$). Moreover, if $(\beta(t),\mathcal{F}_t,Q)$ is any d-dimensional Brownian motion on some probability space $(E,\mathcal{F},Q)$ and if X: $[0,\infty)\times E \longrightarrow \mathbb{R}^N$ is a right continuous, $\{\mathcal{F}_t\}$-progressively measurable function for which (2.7) holds Q-almost surely, then $X(\cdot) = \Phi_{s,x}(\cdot,\beta(*))$ (a.s.,Q) on $\{\xi\in E$: $\beta(*,\xi)$ is continuous$\}$. In particular, $Q\circ X^{-1} = \mathcal{W}\circ\Phi_{s,x}^{-1}$.

<u>Proof</u>: We may and will assume that s = 0 and that x = 0.

In order to construct the mapping $\Phi = \Phi_{0,0}$ on $(\Omega,\mathcal{M},\mathcal{W})$, we begin by setting $\Phi_0(\cdot) \equiv 0$, and, for $n \geq 1$, we define Φ_n inductively by:

$$\Phi_n(T) = \int_0^T \sigma(t,\Phi_{n-1}(t))dx(t) + \int_0^T b(t,\Phi_{n-1}(t))dt, \ T \geq 0.$$

Set $\Delta_n(T) = \sup_{0\leq t\leq T} |\Phi_n(t) - \Phi_{n-1}(t)|$ for $T \geq 0$, and observe that:

$$E^{\mathcal{W}}[\Delta_1(T)^2] \leq 2E^{\mathcal{W}}\left[\sup_{0\leq t\leq T}\left|\int_0^t \sigma(u,0)dx(u)\right|^2\right] + 2E^{\mathcal{W}}\left[\sup_{0\leq t\leq T}\left|\int_0^t b(u,0)du\right|^2\right]$$

$$\leq 8E^{\mathcal{W}}\left[\left|\int_0^T \sigma(u,0)dx(u)\right|^2\right] + 2E^{\mathcal{W}}\left[T\int_0^T |b(u,0)|^2du\right]$$

$$\leq 8E^{\mathcal{W}}\left[\int_0^T \|\sigma(t,0)\|_{H.S.}^2 dt\right] + 2T^2C(T)^2 \leq (8+2T)C(T)^2T.$$

Similarly:

$$E^{\mathcal{W}}[\Delta_{n+1}(T)^2] \leq 8E^{\mathcal{W}}\left[\int_0^T \|\sigma(t,\Phi_n(t)) - \sigma(t,\Phi_{n-1}(t))\|_{H.S.}^2 dt\right] + 2TE^{\mathcal{W}}\left[\int_0^T |b(t,\Phi_n(t)) - b(t,\Phi_{n-1}(t))|^2dt\right]$$

$$\leq (8 + 2T)C(T)^2\int_0^T E^{\mathcal{W}}[|\Phi_n(t) - \Phi_{n-1}(t)|^2dt]$$

$$\leq (8 + 2T)C(T)^2\int_0^T E^{\mathcal{W}}[\Delta_n(t)^2]dt.$$

Hence, by induction on $n \geq 1$, $E^{\mathcal{W}}[\Delta_n(T)^2] \leq K(T)^n/n!$, where $K(T) \equiv (8 + 2T)C(T)^2$; and so:

$$\sup_{n\geq m} E^{\mathcal{W}}\left[\sup_{0\leq t\leq T} |\Phi_n(t) - \Phi_m(t)|^2\right] \leq \sum_{\mu,\nu=m+1}^{\infty} E^{\mathcal{W}}[\Delta_\nu(T)\Delta_\mu(T)]$$

$$\leq \left[\sum_{\nu=m+1}^{\infty} (K(T)^\nu/\nu!)^{1/2}\right]^2 \longrightarrow 0$$

as $m \longrightarrow \infty$. We can therefore find (cf. Lemma (II.2.25)) a right continuous, $\mathcal{W}$-almost surely continuous, $\{\mathcal{M}_t\}$-progressively measurable Φ such that, for all $T > 0$, $E^{\mathcal{W}}\left[\sup_{0\le t\le T}|\Phi(t) - \Phi_n(t)|^2\right] \longrightarrow 0$ as $n \longrightarrow \infty$. In particular,

$$\Phi(T) = \int_0^T \sigma(t,\Phi(t))dx(t) + \int_0^T b(t,\Phi(t))dt,\ T \ge 0,\ (\text{a.s.},\mathcal{W})$$

Finally, suppose that a Brownnian motion $(\beta(t),\mathcal{F}_t,Q)$ on $(E,\mathcal{F},Q)$ and a right continuous, $\{\mathcal{F}_t\}$-progressively measurable solution X to (2.7) are given. Without loss in generality, we assume that $\beta(*,\xi)$ is continuous for all $\xi \in E$. Set $Y(\cdot,\xi) = \Phi(\cdot,\beta(*,\xi))$. Then, as a consequence of ii) in exercise (2.4) and the fact that $Q\circ\beta^{-1} = \mathcal{W}$:

$$Y(T) = \int_0^T \sigma(t,Y(t))d\beta(t) + \int_0^T b(t,Y(t))dt,\ T \ge 0,\ (\text{a.s.},Q).$$

Hence, proceeding in precisely the same way as we did above, we arrive at:

$$E^Q\left[\sup_{0\le t\le T}|X(t) - Y(t)|^2\right] \le K(T)\int_0^T E^Q\left[\sup_{0\le u\le t}|X(u) - Y(u)|^2\right]dt;$$

from which it is easy to conclude that $X(\cdot) = Y(\cdot)$ (a.s.,Q).

Q.E.D.

(2.11) <u>Corollary</u>: Let σ and b be as in the statement of Theorem (2.9), set $a = \sigma\sigma^\dagger$, and define $t \longmapsto L_t$ accordingly. Then, for each $(s,x) \in [0,\infty)\times\mathbb{R}^N$ there is precisely one $P_{s,x} \in M.P.((s,x);\{L_t\})$. In fact, if $\Phi_{s,x}$ is the map described in Theorem (2.9), then $P_{s,x} = \mathcal{W}\circ\Phi_{s,x}^{-1}$.

<u>Proof</u>: Note th t, by Itô's formula, for any $\varphi \in C_0^\infty(\mathbb{R}^N)$:

$$\varphi(\Phi_{s,x}(T)) = \varphi(x) + \int_0^T \sigma^\dagger(s+t,\Phi_{s,x}(t))\nabla\varphi(\Phi_{s,x}(t))dx(t)$$
$$+ \int^T [L_{s+t}\varphi](\Phi_{s,x}(t))dt,\ T \ge 0,$$

(a.s.,$\mathscr{W}$). Hence, $\mathscr{W}\circ\Phi_{s,x}^{-1} \in$ M.P.((s,x);$\{L_t\}$). On the other hand, if P $\in$ M.P.((s,x);$\{L_t\}$), then, by Theorem (2.6), there is a d-dimensional Brownian motion $(\beta(t),\mathscr{F}_t,Q)$ on some probability space $(E,\mathscr{F},Q)$ and a right continuous, $\{\mathscr{F}_t\}$-progressively measurable function $X: [0,\infty)\times E \longrightarrow \mathbb{R}^N$ with the properties that $P = Q\circ X^{-1}$ and (2.7) holds (a.s.,Q). But, by Theorem (2.9), $X(\cdot) = \Phi_{s,x}(\cdot,\beta(*))$ (a.s.,Q), and so $P = Q\circ X^{-1} = \mathscr{W}\circ\Phi_{s,x}^{-1}$.

Q.E.D.

The hypotheses in these results involve σ and b but directly not a. Furthermore, it is clear that when a can be singular, no σ satisfying $a = \sigma\sigma^{\dagger}$ need be as smooth as a is itself (e.g. when N = 1 and $a(x) = |x|$, there is no choice of σ which is Lipschitz continuous at 0, even thought a itself is). The next theorem addresses this problem. In this theorem, and throughout, $a^{1/2}$ denotes the unique $\sigma \in S^+(\mathbb{R}^N)$ satisfying $a = \sigma^2$.

(2.12) <u>Theorem</u>: Given $0 < \epsilon < 1$, set $S_\epsilon^+(\mathbb{R}^N) = \{a\in S^+(\mathbb{R}^N): \epsilon I < a < \frac{1}{\epsilon}I\}$. Then, for $a \in S_\epsilon^+(\mathbb{R}^N)$:

$$a^{1/2} = (1/\epsilon)^{1/2} \sum_{n=0}^{\infty} \begin{bmatrix} 1/2 \\ n \end{bmatrix} (\epsilon a - I)^n, \tag{2.13}$$

where $\begin{bmatrix} 1/2 \\ n \end{bmatrix}$ is the coefficient of z^n in the Taylor series expansion of $(1 + z)^{1/2}$ in $|z| < 1$. Hence, $a \in S_\epsilon^+(\mathbb{R}^N) \longmapsto a^{1/2}$ has bounded derivatives of every order. Moreover, for each $a \in S^+(\mathbb{R}^N)$, $a^{1/2} = \lim_{\epsilon\downarrow 0} (a+\epsilon I)^{1/2}$, and so $a \in S^+(\mathbb{R}^N) \longmapsto a^{1/2}$ is a measurable map. Next, suppose that $\xi \in \mathbb{R}^1 \longmapsto a(\xi) \in S^+(\mathbb{R}^N)$ is a map which satisfies $\mathbf{a}(\xi) \geq \epsilon I$ and $\|\mathbf{a}(\eta) - \mathbf{a}(\xi)\|_{H.S.} \leq C|\eta - \xi|$ for some $\epsilon > 0$ and $C < \infty$ and for all $\xi,\eta \in \mathbb{R}^1$. Then

$$\|\mathbf{a}^{1/2}(\eta) - \mathbf{a}^{1/2}(\xi)\|_{H.S.} \leq C|\eta - \xi|/2\epsilon^{1/2} \tag{2.14}$$

for all $\xi,\eta \in \mathbb{R}^1$. Finally, suppose that $\xi \in \mathbb{R}^1 \longmapsto \mathbf{a}(\xi) \in S^+(\mathbb{R}^N)$ is a twice continuously differentiable map and that $\|\mathbf{a}''(\xi)\|_{op} \leq C < \infty$ for all $\xi \in \mathbb{R}^1$. Then

$$\|\mathbf{a}^{1/2}(\eta) - \mathbf{a}^{1/2}(\xi)\|_{H.S.} \leq N(C)^{1/2}|\eta - \xi| \tag{2.15}$$

for all $\xi,\eta \in \mathbb{R}^1$.

<u>Proof</u>: Equation (2.13) is a simple application of the spectral theorem, as is the equation $\mathbf{a}^{1/2} = \lim_{\epsilon \downarrow 0} (\mathbf{a}+\epsilon I)^{1/2}$. From these, it is clear that $a \in S^+_\epsilon(\mathbb{R}^N) \longrightarrow \mathbf{a}^{1/2}$ has the asserted regularity properties and that $\mathbf{a} \in S^+(\mathbb{R}^N) \longrightarrow \mathbf{a}^{1/2}$ is measurable.

In proving (2.14), we assume, without loss in generality, that $\xi \longrightarrow \mathbf{a}(\xi)$ is continuously differentiable and that $\|\mathbf{a}'(\xi)\|_{H.S.} \leq C$ for all $\xi \in \mathbb{R}^1$; and we will show that, for each $\xi \in \mathbb{R}^1$:

$$\|(\mathbf{a}^{1/2})'(\xi)\|_{H.S.} \leq (1/2\epsilon^{1/2})\|\mathbf{a}(\xi)\|_{H.S.}. \tag{2.16}$$

In proving (2.16), we may and will assume that $\mathbf{a}(\xi)$ is diagonal at ξ. Noting that $\mathbf{a}(\eta) = \mathbf{a}^{1/2}(\eta)\mathbf{a}^{1/2}(\eta)$, we see that:

$$((\mathbf{a}^{1/2})'(\xi))^{ij} = \frac{(a'(\xi))^{ij}}{a^{1/2}(\xi)^{ii} + a^{1/2}(\xi)^{jj}}; \tag{2.17}$$

and clearly (2.16) follows from this.

Finally, to prove (2.15), we will show that if $\mathbf{a}(\xi) > 0$, then $\|((\mathbf{a}^{1/2})'(\xi))\|_{H.S.} \leq N(C)^{1/2}$; and obviously (2.15) follows from this, after an easy limit procedure. Moreover, we will again assume that $\mathbf{a}(\xi)$ is diagnal. But, in this case, it is clear, from (2.17), that we need only check that:

$$|\mathbf{a}'(\xi)^{ij}| \le (C)^{1/2}(a^{ii}(\xi) + a^{jj}(\xi))^{1/2}. \tag{2.18}$$

To this end, set $\varphi_{\pm}(\eta) = \frac{1}{4}(e_i \pm e_j, \mathbf{a}(\eta)(e_i \pm e_j))_{R^N}$ ($(e_1,\ldots,e_N)$ is the standard basis in $\mathbb{R}^N$), and note that $\mathbf{a}'(\xi)^{ij} = \varphi_+'(\xi) - \varphi_-'(\xi)$. Hence, (2.18) will follow once we show that $|\varphi_{\pm}'(\xi)| \le (C)^{1/2}/2$; and this, in turn, will follow if we show that for any $\varphi \in C^2(\mathbb{R}^1)^+$ satisfying $|\varphi''(\eta)| \le K$, $\eta \in \mathbb{R}^1$, $\varphi'(\xi)^2 \le 2K\varphi(\xi)$. But, for any such φ, $0 \le \varphi(\eta) \le \varphi(\xi) + \varphi'(\xi)(\eta-\xi) + 1/2K(\eta-\xi)^2$ for all $\xi,\eta \in \mathbb{R}^1$; and so, by the elementary theory of quadratic inequalities, $\varphi'(\xi)^2 \le 2K\varphi(\xi)$.

Q.E.D.

In view of the preceding results, we have now proved the following existence and uniqueness result for martingale problems.

(2.19) <u>Theorem</u>: Let $\mathbf{a}$: $[0,\infty)\times\mathbb{R}^N \longrightarrow S^+(\mathbb{R}^N)$ and b: $[0,\infty)\times\mathbb{R}^N \longrightarrow \mathbb{R}^N$ be bounded measurable functions. Assume that there is a $C < \infty$ such that

$$|b(t,y) - b(t,y')| \le C|y - y'| \tag{2.20}$$

for all $t \ge 0$ and $y,y' \in \mathbb{R}^N$, and that either

$$\mathbf{a}(t,y) \ge \epsilon I \text{ and } |\mathbf{a}(t,y) - \mathbf{a}(t,y')| \le C|y - y'| \tag{2.21}$$

for all $t \ge 0$ and $y,y' \in \mathbb{R}^N$ and some $\epsilon > 0$ or that $y \in \mathbb{R}^N \longmapsto \mathbf{a}(t,y)$ is twice continuously differentiable for each $t \ge 0$ and that

$$\max_{0\le i\le N}\|\partial^2_{y^i}\mathbf{a}(t,y)\|_{op} \le C,\ (t,y) \in [0,\infty)\times\mathbb{R}^N. \tag{2.22}$$

Let $t \longmapsto L_t$ be the operator-valued map determined by $\mathbf{a}$ and b.

Then, for each $(s,x) \in [0,\infty)\times\mathbb{R}^N$ there is precisely one $P_{s,x} \in$ M.P.$((s,x);\{L_t\})$. In particular, $(s,x)\longmapsto P_{s,x}$ is continuous and, for all stopping times τ, $P_{s,x} = P_{s,x}\otimes_{\tau}P_{\tau,x(\tau)}$.

(2.23) Remark: The preceding cannot be considered to be a particularly interesting application of stochastic integral equations to the study of martingale problems. Indeed, it does little more than give us an alternative (more probabilistic) derivation of the results in section 1 of chapter I. For a more in depth look at this topic, the interested reader should consult Chapter 8 of [S.&V.] where the subject is given a thorough treatment.

3. Localization:

Thus far the hypotheses which we have had to make about a and b in order to check uniqueness are global in nature. The purpose of this section is to prove that the problem of checking whether a martingale probelm is well-posed is a local one. The key to our analysis is contained in the following simple lemma.

(3.1)Lemma: Let $a, \hat{a}$: $[0,\infty)\times\mathbb{R}^N \longrightarrow S^+(\mathbb{R}^N)$ and $b, \hat{b}: [0,\infty)\times\mathbb{R}^N \longrightarrow \mathbb{R}^N$ be bounded measurable functions, and let $t \longmapsto L_t$ and $t \longmapsto \hat{L}_t$ be defined accordingly. Assume that the martingale problem for $\{\hat{L}_t\}$ is well-posed, and denote by $\{\hat{P}_{s,x}: (s,x)\in[0,\infty)\times\mathbb{R}^N\}$ the corresponding family of solutions. If $\mathscr{G}$ is an open subset of $[0,\infty)\times\mathbb{R}^N$ on which a coincides with $\hat{a}$ and b coincides with $\hat{b}$, then, for every $(s,x) \in \mathscr{G}$ and $P \in$ M.P.$((s,x);\{L_t\})$, $P\restriction\mathcal{M}_\zeta = \hat{P}_{s,x}\restriction\mathcal{M}_\zeta$, where $\zeta = \inf\{t\geq 0 : x(t) \notin \mathscr{G}\}$.

Proof: Set $Q = P\otimes_\zeta \hat{P}_{\zeta,x(\zeta)}$. Then, by Theorem (1.18), $Q \in$ M.P.$((s,x);\{\hat{L}_t\})$. Hence, by uniqueness, $Q = \hat{P}_{s,x}$; and so $P\restriction\mathcal{M}_\zeta = Q\restriction\mathcal{M}_\zeta = \hat{P}_{s,x}\restriction\mathcal{M}_\zeta$. Q.E.D.

Given bounded measurable a: $[0,\infty)\times\mathbb{R}^N \longrightarrow S^+(\mathbb{R}^N)$ and b: $[0,\infty)\times\mathbb{R}^N \longrightarrow \mathbb{R}^N$ and the associated map $t \longmapsto L_t$, we say that the martingale problem for $\{L_t\}$ is <u>locally well-posed</u> if $[0,\infty)\times\mathbb{R}^N$ can be covered by open sets U with the property that there exist bounded measurable a_U: $[0,\infty)\times\mathbb{R}^N \longrightarrow S^+(\mathbb{R}^N)$ and b_U: $[0,\infty)\times\mathbb{R}^N \longrightarrow \mathbb{R}^N$ such that $a|_U$ and $b|_U$ coincide with $a_U|_U$ and $b_U|_U$, respectively, and the martingale problem associated with a_U and b_U is well-posed. Our goal is to prove that, under these circumstances, the martingale problem for $\{L_t\}$ is itself well-posed. Thus, until further notice, we will be assuming that the martingale problem for $\{L_t\}$ is locally well-posed, and we will be attempting to prove that there is exactly one element of M.P.$((0,0);\{L_t\})$. The following lemma is a standard application of the Heine-Borel theorem.

(3.2) Lemma: There is a sequence $\{(a_\ell, b_\ell, U_\ell): \ell \in Z^+\}$ such that:

i) $\{U_\ell: \ell \in Z^+\}$ is a locally finite cover of $[0,\infty)\times\mathbb{R}^N$;

ii) for each $\ell \in Z^+$, $a_\ell: [0,\infty)\times\mathbb{R}^N \longrightarrow S^+(\mathbb{R}^N)$ and $b_\ell: [0,\infty)\times\mathbb{R}^N \longrightarrow \mathbb{R}^N$ are bounded measurable functions for which the associated martingale problem is well-posed;

iii) for each $m \in Z^+$ there is an $\epsilon_m > 0$ with the property that whenever $(s,x) \in [m-1,m)\times(B(0,m)\backslash B(0,m-1))$ there exists an $\ell \in Z^+$ for which $[s,s+\epsilon_m]\times\overline{B(x,\epsilon_m)} \subseteq U_\ell$.

Referring to the notation introduced in Lemma (3.2), define $\ell(s,x) = \min\{\ell \in Z^+: [s,s+\epsilon_m]\times\overline{B(x,\epsilon_m)} \subseteq U_\ell\}$ if $(s,x) \in [m-1,m)\times(B(0,m)\backslash B(0,m-1))$. Next, define stopping times σ_n, $n \geq 0$, inductively so that: $\sigma_0 \equiv 0$; $\sigma_n(\omega) = \infty$ if $\sigma_{n-1}(\omega) = \infty$; and, if $\sigma_n(\omega) < \infty$, then $\sigma_n(\omega) = \inf\{t \geq \sigma_{n-1}(\omega): (t,x(t,\omega)) \notin U_{\ell_{n-1}(\omega)}\}$, where $\ell_{n-1}(\omega) = \ell(\sigma_{n-1}(\omega), x(\sigma_n(\omega),\omega))$. Finally, define: $Q^0 = \delta_{\omega_0}$, where $x(\cdot,\omega_0) \equiv 0$; and $Q^{n+1} = [Q^n \otimes_{\sigma_n} P^n_\cdot]\circ(x(\cdot\wedge\sigma_{n+1}))^{-1}$, where $P^n_\omega \equiv P^{\ell_n(\omega)}_{\sigma_n(\omega),x(\sigma_n(\omega),\omega)}$, if $\sigma_n(\omega) < \infty$.

(3.3) Lemma: For each $n \geq 0$ and all $\varphi \in C_0^\infty(\mathbb{R}^N)$:

$$\left(\varphi(x(t)) - \int_0^t [L_u^n \varphi](x(u))du, \mathcal{M}_t, Q^n\right)$$

is a martingale, where $L_t^n \equiv \chi_{[0,\sigma_n)}(t) L_t$, $t \geq 0$.

Proof: We work by induction on $n \geq 0$. Clearly there is nothing to prove when $n = 0$. Now assume the assertion for n, and note that, by Theorem (1.18), we will know it holds for n+1 as soon as we show that for each $\omega \in \{\sigma_n < \infty\}$ and $\delta_\omega \otimes_{\sigma_n(\omega)} P_\omega^n$-almost every $\omega' \in \Omega$: $a(t,x(t,\omega')) = a_{\ell_n(\omega)}(t,x(t,\omega'))$ and $b(t,x(t,\omega')) = b_{\ell_n(\omega)}(t,x(t,\omega'))$ for $t \in [\sigma_n(\omega),\sigma_{n+1}(\omega'))$. But, if $\omega \in \{\sigma_n < \infty\}$, then for $\delta_\omega \otimes_{\sigma_n(\omega)} P_\omega^n$-almost every $\omega' \in \Omega$: $\sigma_n(\omega') = \sigma_n(\omega)$, $x(\cdot\wedge\sigma_n(\omega'),\omega') = x(\cdot\wedge\sigma_n(\omega),\omega)$, and, therefore, $\sigma_{n+1}(\omega') = \inf\{t \geq \sigma_n(\omega): x(t,\omega') \notin U_{\ell_n(\omega)}\}$. Since a and b coincide with $a_{\ell_n(\omega)}$ and $b_{\ell_n(\omega)}$, respectively, on $U_{\ell_n(\omega)}$ whenever $\sigma_n(\omega) < \infty$, the proof is complete.

Q.E.D.

(3.4) Lemma: For each $T > 0$, $\lim_{t\to\infty} Q^n(\sigma_n \leq T) = 0$. In particular, there is a unique $Q \in \mathbf{M}_1(\Omega)$ such that $Q\restriction\mathcal{M}_{\sigma_n} = Q^n\restriction\mathcal{M}_{\sigma_n}$ for all $n \geq 0$. Finally, $Q \in$ M.P.$((0,0);\{L_t\})$.

Proof: By Lemma (3.3) and exercise (1.9), there exists for each $T > 0$ a $C(T) < \infty$ (depending only on the bounds on a and b) such that $E^{Q^n}\left[|x(t) - x(s)|^4\right] \leq C(T)(t - s)^2$, $0 \leq s < t \leq T$. Hence, since $Q^n(x(0)=0) = 1$ for all $n \geq 0$, $\{Q^n: n \geq 0\}$ is relatively compact in $\mathbf{M}_1(\Omega)$. Next, note that $\sigma_n(\omega)\uparrow\infty$

uniformly fast for ω's in a compact subset of Ω. Hence, $Q^n(\sigma_n \le T) \longrightarrow 0$ as $n \longrightarrow \infty$ for each $T > 0$. Also, observe that $Q^n \restriction \mathcal{M}_{\sigma_m} = Q^m \restriction \mathcal{M}_{\sigma_m}$ for all $0 \le m \le n$. Combining this with the preceding, we now see that for all $T > 0$ and all $\Gamma \in \mathcal{M}_T$: $\lim_{n \longrightarrow \infty} Q^n(\Gamma)$ exists. Thus, $\{Q^n\}$ can has precisely one limit Q; and clearly $Q \restriction \mathcal{M}_{\sigma_n} = Q^n \restriction \mathcal{M}_{\sigma_n}$ for each $n \ge 0$. Finally, we see from Lemma (3.3) that

$$E^Q[\varphi(x(t\wedge\sigma_n)) - \varphi(x(s\wedge\sigma_n), \Gamma] = E^Q\left[\int_{s\wedge\sigma_n}^{t\wedge\sigma_n}[L_u\varphi](x(u))du, \Gamma\right]$$

for all $n \ge 0$, $\varphi \in C_0^\infty(\mathbb{R}^N)$, $0 \le s < t$, and $\Gamma \in \mathcal{M}_s$; and therefore $Q \in M.P.((0,0);\{L_t\})$. Q.E.D.

(3.5) <u>Theorem</u>: Let $a:[0,\infty)\times\mathbb{R}^N \longrightarrow S^+(\mathbb{R}^N)$ and $b: [0,\infty)\times\mathbb{R}^N \longrightarrow \mathbb{R}^N$ be bounded measurable functions and define $t \longmapsto L_t$ accordingly. If the martingale problem for $\{L_t\}$ is locally well-posed, then it is in fact well-posed.

<u>Proof</u>: Clearly, it is sufficient for us to prove that $M.P.((0,0);\{L_t\})$ contains precisely one element. Since we already know that the Q constructed in Lemma (3.4) is one element of $M.P.((0,0);\{L_t\})$, it remains to show that it is the only one. To this end, suppose that P is a second one. Then, by Lemma (3.1), $P \restriction \mathcal{M}_{\sigma_1} = P_{0,0}^{\ell(0,0)} \restriction \mathcal{M}_{\sigma_0} = Q \restriction \mathcal{M}_{\sigma_0}$. Next, assume that $P \restriction \mathcal{M}_{\sigma_n} = Q \restriction \mathcal{M}_{\sigma_n}$ and let $\omega \longmapsto P_\omega$ be a r.c.p.d. of $P|\mathcal{M}_{\sigma_n}$. Then, for P-almost every $\omega \in \{\sigma_n < \infty\}$, $P_\omega \circ \theta_{\sigma_n}^{-1} \in M.P.((\sigma_n(\omega), x(\sigma_n(\omega),\omega);\{L_t\})$ and therefore, by Lemma (3.1),

$P_\omega \circ \theta^{-1}_{\sigma_n(\omega)} \restriction \mathcal{M}_{\zeta_\omega} = P^{\ell_n(\omega)}_{\sigma_n(\omega), x(\sigma_n(\omega), x(\sigma_n(\omega),\omega)} \restriction \mathcal{M}_{\zeta_\omega}$, where $\zeta_\omega(\omega') = \inf\{t \geq 0: (t, x(t,\omega')) \notin U_{\ell_n(\omega)}\}$. (Recall that $\theta_t : \Omega \longmapsto \Omega$ is the time shift map.) At the same time, if $\sigma_n(\omega) < \infty$, then $\sigma_{n+1}(\omega') = \sigma_n(\omega) + \zeta_\omega(\theta_{\sigma_n(\omega)}\omega')$ for P_ω-almost every ω'. Combining these, we see that

$$P_\omega \restriction \mathcal{M}_{\sigma_{n+1}} = \delta_\omega \otimes_{\sigma_n(\omega)} P^{\ell_n(\omega)}_{\sigma_n(\omega), x(\sigma_n(\omega), x(\sigma_n(\omega),\omega)} \restriction \mathcal{M}_{\sigma_{n+1}}$$

for P-almost every $\omega \in \{\sigma_n < \infty\}$. Since, by induction hypothesis, we already know that $P \restriction \mathcal{M}_{\sigma_n} = Q \restriction \mathcal{M}_{\sigma_n}$, we can now conclude that $P \restriction \mathcal{M}_{\sigma_{n+1}} = Q \restriction \mathcal{M}_{\sigma_{n+1}}$. Because $\sigma_n \uparrow \infty$ (a.s.,Q), we have therefore proved that $P = Q$.

Q.E.D.

The following is a somewhat trivial application of the preceding. We will have a much more interesting one in section 5 below.

(3.6) <u>Corollary</u>: Suppose that $a: [0,\infty) \times \mathbb{R}^N \longrightarrow S^+(\mathbb{R}^N)$ and $b: [0,\infty) \times \mathbb{R}^N \longrightarrow \mathbb{R}^N$ are bounded measurable functions with the properties that $a(t,\cdot)$ has two continuous derivatives and $b(t,\cdot)$ has one continuous derivative for each $t \geq 0$. Further, assume that the second derivatives of $a(t,\cdot)$ and the first derivatives of $b(t,\cdot)$ at $x = 0$ are uniformly bounded for t in compact intervals. Then, the martingale problem for the associated $\{L_t\}$ is well-posed and the corresponding family $\{P_{s,x}: (s,x) \in [0,\infty) \times \mathbb{R}^N\}$ is continuous.

4. The Cameron-Martin-Girsanov Transformation:

It is clear on analytic grounds that if the coefficient matrix a is strictly positive definite then the first order part of the operator L_t is a lower order perturbation away from its principle part

$$L_t^0 = 1/2 \sum_{i,j=1}^{N} a^{ij}(t,y)\partial_{y^i}\partial_{y^j} \tag{4.1}$$

Hence, one should suspect that, in this case, the martingale problems corresponding to $\{L_t^0\}$ and $\{L_t\}$ are closely related. In this section we will confirm this suspicion. Namely, we are going to show that when a is uniformly positive definite, then, at least over finite time intervals, P's in M.P.$((s,x);\{L_t\})$ differ from P's in M.P.$((s,x);\{L_t^0\})$ by a quite explicit Radon-Nikodym factor.

(4.2)Lemma: Let $(R(t),\mathcal{M}_t,P)$ be a non-negative martingale with $R(0) \equiv 1$. Then there is a unique $Q \in \mathbf{M}_1(\Omega)$ such that $Q\restriction\mathcal{M}_T = R(T)P\restriction\mathcal{M}_T$ for each $T \geq 0$.

Proof: The uniqueness assertion is obvious. To prove the existence, define $Q^n = (R(n)P)\circ(x(\cdot\wedge n))^{-1}$ for $n \geq 0$. Then $Q^{n+1}\restriction\mathcal{M}_n = Q^n\restriction\mathcal{M}_n$; from which it is clear that $\{Q^n\colon n\geq 0\}$ is relatively compact in $\mathbf{M}_1(\Omega)$. In addition, one sees that any limit of $\{Q^n\colon n\geq 0\}$ must have the required property.

Q.E.D.

(4.3)Lemma: Let $(R(t),\mathcal{M}_t,P)$ be an P-almost surely continuous strictly positive martingale satisfying $R(0) \equiv 1$. Define $Q \in \mathbf{M}_1(\Omega)$ accordingly as in Lemma (4.2) and set $\mathcal{R} =$

logR. Then $(1/R(t), \mathcal{M}_t, Q)$ is a Q-almost surely continuous strictly positive martingale, and $P\restriction\mathcal{M}_T = (1/R(T))Q\restriction\mathcal{M}_T$ for all $T \geq 0$. Moreover, $\mathcal{R} \in S.Mart_c(\{\mathcal{M}_t\}, P)$,

$$\mathcal{R}(T) = \int_0^T \frac{1}{R(t)} dR(t) - \frac{1}{2}\int_0^T \left(\frac{1}{R(t)}\right)^2 \langle R, R\rangle(dt) \tag{4.4}$$

(a.s.,P) for $T \geq 0$; and $X \in Mart_c^{loc}(\{\mathcal{M}_t\}, P)$ if and only if $X^R \equiv X - \langle X, \mathcal{R}\rangle \in Mart_c^{loc}(\{\mathcal{M}_t\}, Q)$. In particular, $S.Mart_c(\{\mathcal{M}_t\}, P) = S.Mart_c(\{\mathcal{M}_t\}, Q)$. Finally, if $X, Y \in S.Mart_c(\{\mathcal{M}_t\}, P)$, then, up to a P,Q-null set, $\langle X, Y\rangle$ is the same whether it is computed relative to $(\{\mathcal{M}_t\}, P)$ or to $(\{\mathcal{M}_t\}, Q)$. In particular, given $X \in S.Mart_c(\{\mathcal{M}_t\}, P)$ and an $\{\mathcal{M}_t\}$-progressively measurable $\alpha: [0,\infty)\times\Omega \longrightarrow \mathbb{R}^1$ satisfying $\int_0^T \alpha(t)\langle X, X\rangle(dt) < \infty$ (a.s.,P) for all $T > 0$, the quantity $\int_0^{\cdot} \alpha dX$ is, up to a P,Q-null set, is the same whether it is computed relative to $(\{\mathcal{M}_t\}, P)$ or to $(\{\mathcal{M}_t\}, Q)$.

<u>Proof</u>: The first assertion requiring comment is that (4.4) holds; from which it is immediate that $\mathcal{R} \in S.Mart_c(\{\mathcal{M}_t\}, P)$. But applying Itô's formula to $\log(R(t)+\epsilon)$ for $\epsilon > 0$ and then letting $\epsilon \downarrow 0$, we obtain (4.4) in the limit. In proving that $X \in Mart_c^{loc}(\{\mathcal{M}_t\}, P)$ implies that $X^R \in Mart_c^{loc}(\{\mathcal{M}_t\}, Q)$, we may and will assume that R, 1/R, and X are all bounded. Given $0 \leq t_1 < t_2$ and $A \in \mathcal{M}_{t_1}$, we have:

$$\begin{aligned} E^Q\left[X(t_2) - \frac{1}{R(t_2)}\langle X, X\rangle(t_2), A\right] &= E^P\left[R(t_2)X(t_2) - \langle X, R\rangle(t_2), A\right] \\ &= E^P\left[R(t_1)X(t_1) - \langle X, R\rangle(t_1), A\right] \\ &= E^Q\left[X(t_1) - \frac{1}{R(t_1)}\langle X, X\rangle(t_1), A\right]. \end{aligned}$$

Hence, since, by (4.4), $\langle X,\mathcal{R}\rangle(dt) = \frac{1}{R(t)}\langle X,X\rangle(dt)$, we will be done once we show that $\frac{1}{R(\cdot)}\langle X,X\rangle(\cdot) - \int_0^{\cdot}\frac{1}{R(s)}\langle X,X\rangle(ds) \in \mathrm{Mart}_c^{loc}(\{\mathcal{M}_t\},Q)$. However, by Itô's formula

$$\frac{1}{R(T)}\langle X,X\rangle(T) = \int_0^T\langle X,X\rangle(t)d\left[\frac{1}{R(t)}\right] + \int_0^T\frac{1}{R(t)}\langle X,X\rangle(dt),$$

and so the required conclusion follows from the fact that $(1/R(t),\mathcal{M}_t,Q)$ is a martingale. We have now shown that $X^R \in \mathrm{Mart}_c^{loc}(\{\mathcal{M}_t\},Q)$ whenever $X \in \mathrm{Mart}_c^{loc}(\{\mathcal{M}_t\},P)$ and therefore that $S.\mathrm{Mart}_c(\{\mathcal{M}_t\},P) \subseteq S.\mathrm{Mart}_c(\{\mathcal{M}_t\},Q)$. Because the roles of P and Q are symmetric, we will have proved the opposite implications as soon as we show that $\langle X,Y\rangle$ is the same under $(\{\mathcal{M}_t\},P)$ and $(\{\mathcal{M}_t\},Q)$ for all $X,Y \in S.\mathrm{Mart}_c(\{\mathcal{M}_t\},P)$.

Finally, let $X,Y \in \mathrm{Mart}_c^{loc}(\{\mathcal{M}_t\},P)$. To see that $\langle X,Y\rangle$ is the same for $(\{\mathcal{M}_t\},P)$ and $(\{\mathcal{M}_t\},Q)$, we must show that $X^RY^R - \langle X,Y\rangle_P \in \mathrm{Mart}_c^{loc}(\{\mathcal{M}_t\},Q)$ (where the subscript P is used to emphasize that $\langle X,Y\rangle$ has been computed relative to $(\{\mathcal{M}_t\},P)$). However, by Itô's formula:

$$X^RY^R(T) = XY(0) + \int_0^T X^R(t)dY^R(t)$$
$$+ \int_0^T Y^R(t)dX^R(t) + \langle X^R,Y^R\rangle_P(T),\ T \geq 0,$$

(a.s.,P). Thus, it remains to check that $\int_0^{\cdot}X^RdY^R$ and $\int_0^{\cdot}Y^RdX^R$ are elements of $\mathrm{Mart}_c^{loc}(\{\mathcal{M}_t\},Q)$. But:

$$\int_0^{\cdot}X^RdY^R = \int_0^{\cdot}X^RdY - \int_0^{\cdot}X^Rd\langle Y,\mathcal{R}\rangle_P = \int_0^{\cdot}X^RdY - \langle\int_0^{*}X^RdY,\mathcal{R}\rangle(\cdot) =$$
$$= \left[\int_0^{\cdot}X^RdY\right]^R \in \mathrm{Mart}_c^{loc}(\{\mathcal{M}_t\},Q),$$

and, by symmetry, the same is true of $\int_0^{\cdot}Y^RdX^R$. Q.E.D.

(4.5)<u>Exercise</u>: A more constructive proof that $\langle X,Y\rangle$ is the same under P and Q can be based on the observation that $\langle X\rangle(T)$ can be expressed in terms of the quadratic variation of $X(\cdot)$ over $[0,T]$ (cf. exercises (II.2.28) and (II.3.14)).

(4.6)<u>Theorem</u> (Cameron-Martin-Girsanov): Let $a: [0,\infty)\times\mathbb{R}^N \longrightarrow S^+(\mathbb{R}^N)$ and $b,c: [0,\infty)\times\mathbb{R}^N \longrightarrow \mathbb{R}^N$ be bounded measurable functions and let $t\longmapsto L_t$ and $t\longmapsto \tilde{L}_t$ be the operators associated with a and b and with a and b+ac, respectively. Then $Q \in M.P.((s,x);\{\tilde{L}_t\})$ if and only if there is a $P \in M.P.((s,x);\{L_t\})$ such that $Q\restriction\mathcal{M}_T = R(T)P\restriction\mathcal{M}_T$, $T \geq 0$, where

$$(4.7)\qquad R(T) = \exp\left[\int_0^T c(s+t,x(t))d\bar{x}(t) - 1/2\int_0^T (c,ac)_{\mathbb{R}^N}(s+t,x(t))dt\right]$$

with $\bar{x}(T) \equiv x(T) - \int_0^T b(s+t,x(t))dt$, $T \geq 0$. In particular, for each $(s,x) \in [0,\infty)\times\mathbb{R}^N$, there is a one-to-one correspondence between $M.P.((s,x);\{\tilde{L}_t\})$ and $M.P.((s,x);\{L_t\})$.

<u>Proof</u>: Suppose that $P \in M.P.((s,x);\{L_t\})$ and define $R(\cdot)$ by (4.7). By part ii) in exercise (II.3.13), $(R(t),\mathcal{M}_t,P)$ is a martingale; and, clearly, $R(0) \equiv 0$ and $R(\cdot)$ is P-almost surely positive. Thus, by Lemmas (4.2) and (4.3), there is a unique $Q \in M_1(\Omega)$ such that $Q\restriction\mathcal{M}_T = R(T)P\restriction\mathcal{M}_T$, $T \geq 0$. Moreover, $X \in Mart_c^{loc}(\{\mathcal{M}_t\},P)$ if and only if $X - \langle X,\mathfrak{R}\rangle \in Mart_c^{loc}(\{\mathcal{M}_t\},Q)$, where $\mathfrak{R} = \log R$. In particular, since

$$\langle X,\mathfrak{R}\rangle(dt) = \sum_{i=1}^N c_i(s+t,x(t))\langle x^i,X\rangle(dt),$$

if $\varphi \in C_0^\infty(\mathbb{R}^N)$, then

$$\langle\varphi(x(\cdot)),\mathscr{R}\rangle(dt) = \sum_{i=1}^N c_i(s+t,x(t))\langle x^i,\varphi(x(\cdot))\rangle(dt)$$

$$= \sum_{i,j=1}^N (c_i a^{ij}\partial_{x^j}\varphi)(s+t,x(t))dt$$

$$= \sum_{j=1}^N (ac)^j(s+t,x(t))\partial_{x^j}\varphi(x(t))dt,$$

and so $(\varphi(x(t)) - \int_0^t[\tilde{L}_u\varphi](x(u))du,\mathscr{M}_t,Q)$ is a martingale. In other words, $Q \in M.P.((s,x);\{\tilde{L}_t\})$.

Conversely, suppose that $Q \in M.P.((s,x);\{\tilde{L}_t\})$ and define $R(\cdot)$ as in (4.7) relative to Q. Then:

$$\frac{1}{R(T)} = \exp\left[-\int_0^T c(s+t,x(t))d\tilde{x}(t) - 1/2\int_0^T (c,ac)(s+t,x(t))dt\right]$$

where $\tilde{x}(T) \equiv x(T) - \int_0^T (b+ac)(s+t,x(t))dt$. Hence, by the preceding paragraph applied to Q and $\{\tilde{L}_t\}$, we see that there is a unique $P \in M.P.((s,x);\{L_t\})$ such that $P\restriction\mathscr{M}_T = \frac{1}{R(T)}Q\restriction\mathscr{M}_T$, $T \geq 0$. Since stochastic integrals are the same whether they are defined relative to P or Q, we now see that $Q\restriction\mathscr{M}_T = R(T)P\restriction\mathscr{M}_T$, $T \geq 0$, where $R(\cdot)$ is now defined relative to P.

Q.E.D.

(4.8) Corollary: Let a: $[0,\infty)\times\mathbb{R}^N \longrightarrow S^+(\mathbb{R}^N)$ and b: $[0,\infty)\times\mathbb{R}^N \longrightarrow \mathbb{R}^N$ be bounded measurable functions and assume that a is uniformly positive definite on compact subsets of $[0,\infty)\times\mathbb{R}^N$. Define $t\longmapsto L_t^0$ as in (4.1) and let $t\longmapsto L_t$ be the operator associated with a and b. Then, the martingale problem for $\{L_t^0\}$ is well-posed if and only if the martingale problem for $\{L_t\}$ is well-posed.

Proof: In view of Theorem (3.5), we may and will assume that a is uniformly positive definite on the whole of $[0,\infty)\times\mathbb{R}^N$. But we can then take $c = a^{-1}b$ and apply Theorem (4.6).

Q.E.D.

5. The Martingale Problem when a is Continuous and Positive:

Let $a: \mathbb{R}^N \longrightarrow S^+(\mathbb{R}^N)$ be a bounded continuous function satisfying $a(x) > 0$ for each $x \in \mathbb{R}^N$. Let $b: [0,\infty)\times\mathbb{R}^N \longrightarrow \mathbb{R}^N$ be a bounded measurable function. Our goal in this section is to prove that the martingale problem associated with a and b is well-posed.

In view of Corollary (4.8), we may and will assume that $b \equiv 0$, in which case existence presents no problem. Moreover, because of Theorem (3.5), we may and will assume in addition that

$$\|a(x) - I\|_{H.S.} \leq \epsilon, \tag{5.1}$$

where $\epsilon > 0$ is as small as we like.

Set $L = 1/2 \sum_{i,j=1}^{N} a^{ij}(y)\partial_{y^i}\partial_{y^j}$. What we are going to do is show that when the ϵ in (5.1) is sufficiently small then, for each $\lambda > 0$, there is a map S_λ from the Schwartz space $\mathcal{S}(\mathbb{R}^N)$ into $C_b(\mathbb{R}^N)$ such that

$$E^P\left[\int_0^\infty e^{-\lambda t} f(x(t))dt\right] = S_\lambda f(x) \tag{5.2}$$

for all $f \in \mathcal{S}(\mathbb{R}^N)$, whenever $P \in M.P.(x;L)$. Once we have proved (5.2), the argument is easy. Namely, if P and Q are elements of $M.P.(x;L)$, then (5.2) allows us to say that

$$E^P\left[\int_0^\infty e^{-\lambda t} f(x(t))dt\right] = E^Q\left[\int_0^\infty e^{-\lambda t} f(x(t))dt\right]$$

for all $\lambda > 0$ and $f \in C_b(\mathbb{R}^N)$. But, by the uniqueness of the Laplace transform, this means that $P\circ x(t)^{-1} = Q\circ x(t)^{-1}$ for all $t \geq 0$; and so, by Corollary (1.15), $P = Q$. Hence, everything reduces to proving (5.2).

(5.3)<u>Lemma</u>: Set $\gamma_t = g(t,\cdot)$, where $g(t,y)$ denotes the standard Gauss kernel on $\mathbb{R}^N$; and, for $\lambda > 0$, define $R_\lambda f = \int_0^\infty e^{-\lambda t}\gamma_t * f\,dt$ for $f \in \mathcal{S}(\mathbb{R}^N)$ ("$*$" denotes convolution). Then R_λ maps $\mathcal{S}(\mathbb{R}^N)$ into itself and $(\lambda I - \frac{1}{2}\Delta)\circ R_\lambda = R_\lambda \circ (\lambda I - \frac{1}{2}\Delta) = I$ on $\mathcal{S}(\mathbb{R}^N)$. Moreover, if $p \in (N/2,\infty)$, then there is an $A = A(\lambda,p) \in (0,\infty)$ such that

$$\|R_\lambda f\|_{L^p(\mathbb{R}^N)} \le A\|f\|_{L^p(\mathbb{R}^N)}, \quad f \in \mathcal{S}(\mathbb{R}^N). \tag{5.4}$$

Finally, for every $p \in (1,\infty)$ there is a $C = C(p) \in (0,\infty)$ (i.e. independent of $\lambda > 0$) such that

$$\Big\|\Big(\sum_{i,j=1}^N (\partial_{y^i}\partial_{y^j}R_\lambda f)^2\Big)^{1/2}\Big\|_{L^p(\mathbb{R}^N)} \le C\,\|f\|_{L^p(\mathbb{R}^N)} \tag{5.5}$$

for all $f \in \mathcal{S}(\mathbb{R}^N)$.

<u>Proof</u>: Use $\mathcal{F}f$ to denote the Fourier transform of f. Then it is an easy computation to show that $\mathcal{F}R_\lambda f(\xi) = (\lambda + \frac{1}{2}|\xi|^2)^{-1}\mathcal{F}f(\xi)$. From this it is clear that R_λ maps $\mathcal{S}(\mathbb{R}^N)$ into itself and that R_λ is the inverse on $\mathcal{S}(\mathbb{R}^N)$ of $(\lambda I - \frac{1}{2}\Delta)$. To prove the estimate (5.4), note that $\|\gamma_t * f\|_{C_b(\mathbb{R}^N)} \le \|\gamma_t\|_{L^q(\mathbb{R}^N)}\|f\|_{L^p(\mathbb{R}^N)}$, where $1/q = 1 - 1/p$, and that $\|\gamma_t\|_{L^q(\mathbb{R}^N)} = B_N t^{N/(2p)}$ for some $B_N \in (0,\infty)$. Thus, if $p \in (N/2,\infty)$, then

$$\|R_\lambda f\|_{L^p(\mathbb{R}^N)} \le B_N\left[\int_0^\infty e^{-\lambda t}t^{-N/(2p)}dt\right]\|f\|_{L^p(\mathbb{R}^N)} = A\|f\|_{L^p(\mathbb{R}^N)},$$

where $A \in (0,\infty)$.

The estimate (5.5) is considerably more sophisticated.

What it comes down to is the proof that for each $p \in (1,\infty)$ there is a $K = K(p) \in (0,\infty)$ such that

$$\|(\sum_{i,j=1}^{N}(\partial_{y^i}\partial_{y^j}f)^2)^{1/2}\|_{L^p(\mathbb{R}^N)} \le K\|\tfrac{1}{2}\Delta f\|_{L^p(\mathbb{R}^N)} \tag{5.6}$$

for $f \in \mathcal{S}(\mathbb{R}^N)$. Indeed, suppose that (5.6) holds. Then, since $\frac{1}{2}\Delta R_\lambda = I - \lambda R_\lambda$, we would have

$$\|(\sum_{i,j=1}^{N}(\partial_{y^i}\partial_{y^j}R_\lambda f)^2)^{1/2}\|_{L^p(\mathbb{R}^N)} \le K\|\tfrac{1}{2}\Delta R_\lambda f\|_{L^p(\mathbb{R}^N)}$$

$$\le K\|f\|_{L^p(\mathbb{R}^N)} + K\|\lambda R_\lambda f\|_{L^p(\mathbb{R}^N)} \le 2K\|f\|_{L^p(\mathbb{R}^N)},$$

since $\|\gamma_t\|_{L^1(\mathbb{R}^N)} = 1$ and so $\|\lambda R_\lambda f\|_{L^p(\mathbb{R}^N)} \le \|f\|_{L^p(\mathbb{R}^N)}$. Except when $p = 2$, (5.6) has no elementary proof and depends on the theory of singular integral operators. Rather than spend time here developing the relevant theory, we will defer the proof to the appendix which follows this section. Q.E.D.

Choose and fix a $p \in (N/2,\infty)$ and take the ϵ in (5.6) to lie in the interval $(0,\frac{1}{2\sqrt{C(p)}})$, where $C(p)$ is the constant C in (5.5). We can now define the operator S_λ. Namely, set $D_\lambda = (L-\frac{1}{2}\Delta)R_\lambda$. Then, for $f \in \mathcal{S}(\mathbb{R}^N)$:

$$\|D_\lambda f\|_{L^p(\mathbb{R}^N)} = \tfrac{1}{2}\|\sum_{i,j=1}^{N}(a-I)^{ij}\partial_{y^i}\partial_{y^j}R_\lambda f\|_{L^p(\mathbb{R}^N)}$$

$$\le \tfrac{\epsilon}{2}\|(\sum_{i,j=1}^{N}(\partial_{y^i}\partial_{y^j}R_\lambda f)^2)^{1/2}\|_{L^p(\mathbb{R}^N)} \le 1/2\|f\|_{L^p(\mathbb{R}^N)}.$$

Hence, D_λ admits a unique extension as a continuous operator on $L^p(\mathbb{R}^N)$ with bound not exceeding 1/2. Using D_λ again to denote this extension, we see that $I-D_\lambda$ admits an continuous inverse K_λ with bound not larger than 2. We now define $S_\lambda =$

$R_\lambda \circ K_\lambda$.

Note that if $K_\lambda f \in \mathcal{S}(\mathbb{R}^N)$, then $(\lambda I-L)S_\lambda f = (\lambda I-\frac{1}{2}\Delta)R_\lambda K_\lambda f - D_\lambda K_\lambda f = (I-D_\lambda)K_\lambda f = f$. Thus, if $\mathcal{F}_\lambda \equiv \{f \in L^p(\mathbb{R}^N): K_\lambda f \in \mathcal{S}(\mathbb{R}^N)\}$, then we have that:

$$(\lambda I-L)S_\lambda f = f,\ f \in \mathcal{F}_\lambda. \tag{5.7}$$

In particular, we see that $\mathcal{F}_\lambda \subseteq C_b(\mathbb{R}^N)$. Moreover, since $\mathcal{S}(\mathbb{R}^N)$ is dense in $L^p(\mathbb{R}^N)$ and K_λ is invertible, it is clear that $\mathcal{F}_\lambda$ is also dense in $L^p(\mathbb{R}^N)$. We will now show that (5.2) holds for all $P \in M.P.(x;L)$ and $f \in \mathcal{F}_\lambda$. Indeed, if $f \in \mathcal{F}_\lambda$, then an easy application of Itô's formula in conjunction with (5.7) shows that

$$\left(e^{-\lambda t}S_\lambda f(x(t))+\int_0^t e^{-\lambda s}f(x(s))ds,\ \mathcal{M}_t, P\right)$$

is a martingale for every $P \in M.P.(x;L)$. Thus, (5.2) follows by letting $t\longrightarrow\infty$ in

$$E^P[e^{-\lambda t}S_\lambda f(x(t))] - S_\lambda f(x) = E^P\left[\int_0^t e^{-\lambda s}f(x(s))ds\right].$$

At first sight, we appear to be very close to a proof of (5.2) for all $f \in \mathcal{S}(\mathbb{R}^N)$. However, after a little reflection, one will realize that we still have quite a long way to go. Indeed, although we now know that (5.2) holds for all $f \in \mathcal{F}_\lambda$ and that $\mathcal{F}_\lambda$ is a dense subset of $L^p(\mathbb{R}^N)$, we must still go through a limit procedure before we can assert that (5.2) holds for all $f \in \mathcal{S}(\mathbb{R}^N)$. The right hand side of (5.2) presents no problems at this point. In fact, (5.4) says that $f \in L^p(\mathbb{R}^N) \longmapsto R_\lambda f(x)$ is a continuous map for each $x \in \mathbb{R}^N$, and therefore S_λ has the same property. On the other hand, we know nothing about the behavior of the left hand side of

(5.2) under convergence in $L^p(\mathbb{R}^N)$. Thus, in order to complete our program we have still to prove an <u>a priori</u> estimate which says that for each $P \in M.P.(x;L)$ there is a $B \in (0,\infty)$ such that

$$\left|E^P\left[\int_0^\infty e^{-\lambda t} f(x(t))dt\right]\right| \le B\|f\|_{L^p(\mathbb{R}^N)}, \quad f \in \mathcal{S}(\mathbb{R}^N). \tag{5.8}$$

To prove (5.8), let $P \in M.P.(x;L)$ be given. Then, by Theorem (2.6), there is an N-dimensional Brownian motion $(\beta(t),\mathcal{M}_t,P)$ such that $x(T) = x + \int_0^T \sigma(x(t))d\beta(t)$, $T \ge 0$ (a.s.,P), where $\sigma = a^{1/2}$. Set $\sigma_n(t,\omega) = \sigma(x(\frac{[nt]}{n}\wedge n,\omega))$ and $X_n(T) = x + \int_0^T \sigma_n(t)d\beta(t)$, $T \ge 0$. Note that, for each $T \ge 0$,

$$E^P\left[\sup_{0\le t\le T\wedge n}|x(t) - X_n(t)|^2\right]$$
$$\le 4E^P\left[\int_0^T \|\sigma(x(t))-\sigma(x(\tfrac{[nt]}{n}))\|^2_{H.S.}dt\right]\longrightarrow 0$$

as $n\longrightarrow\infty$. Hence, if $\mu_n \in \mathbf{M}_1(\mathbb{R}^N)$ is defined by

$$\int_{\mathbb{R}^N} f d\mu_n = \lambda E^P\left[\int_0^\infty e^{-\lambda t} f(X_n(t))dt\right], \quad f \in C_b(\mathbb{R}^N),$$

then $\mu_n\longrightarrow\mu$ in $\mathbf{M}_1(\mathbb{R}^N)$, where

$$\int_{\mathbb{R}^N} f d\mu = \lambda E^P\left[\int_0^\infty e^{-\lambda t} f(x(t))dt\right], \quad f \in C_b(\mathbb{R}^N).$$

In particular, if

$$\left|\int f d\mu_n\right| \le \lambda B\|f\|_{L^p(\mathbb{R}^N)}, \quad f \in \mathcal{S}(\mathbb{R}^N), \tag{5.9}$$

for some $B \in (0,\infty)$ and all $n \ge 1$, then (5.8) holds for the same B.

(5.10)<u>Lemma</u>: For all $n \ge 1$, the estimate (5.9) holds with $B = 2A$, where $A = A(\lambda,p)$ is the constant in (5.4).

<u>Proof</u>: Choose and fix $n \ge 1$. We will first show that if

(5.9) holds for some $B \in (0,\infty)$, then it holds with $B = 2A$. To this end, note that $X_n \in (\mathrm{Mart}_c^2(\{\mathcal{M}_t\},P))^N$ and that $\langle\langle X_n,X_n\rangle\rangle(dt) = a_n(t)dt$ where $a_n(t,\omega) = a(x(\frac{[nt]}{n}\wedge n,\omega))$. Hence, by Itô's formula, for $f \in \mathcal{S}(\mathbb{R}^N)$:

$$\left(e^{-\lambda t}R_\lambda f(x(t)) + \int_0^t e^{-\lambda s}(f(x(s)-\psi(s))ds,\ \mathcal{M}_t,P\right)$$

is a martingale, where

$$\psi(t,\omega) \equiv 1/2 \sum_{i,j=1}^{N} (a_n(t,\omega)-I)^{ij}\partial_{y^i}\partial_{y^j}R_\lambda f.$$

Hence, we have that

$$\lambda R_\lambda f(x) + \lambda E^P\left[\int_0^\infty e^{-\lambda t}\psi(t)dt\right] = \int f d\mu_n.$$

Noting that $|\psi(t,\omega)| \le \frac{\epsilon}{2}\left[\sum_{i,j=1}^{N} (\partial_{y^i}\partial_{y^j}R_\lambda f(X_n(t,\omega)))^2\right]^{1/2}$, we see that

$$\left|E^P\left[\int_0^\infty e^{-\lambda t}\psi(t)dt\right]\right| \le \frac{\epsilon}{2}\int_{\mathbb{R}^N} \left[\sum_{i,j=1}^{N} (\partial_{y^i}\partial_{y^j}R_\lambda f)^2\right]^{1/2} d\mu_n$$

$$\le \frac{\epsilon CM}{2}\|f\|_{L^p(\mathbb{R}^N)} \le \frac{M}{2}\|f\|_{L^p(\mathbb{R}^N)},$$

where M denotes the smallest B for which (5.9) holds. Using this and (5.4) in the preceding, we obtain $M \le \lambda A + M/2$, from which $M \le 2\lambda A$ is immediate.

We must still check that (5.9) holds for some $B \in (0,\infty)$. Let $f \in \mathcal{S}(\mathbb{R}^N)^+$ be given. Then

$$\int f d\mu_n = \sum_{m=0}^{n^2-1} \lambda e^{-\lambda m/n}E^P\left[\int_0^{1/n} e^{-\lambda t}f(X_n(\tfrac{m}{n}+t))dt\right]$$

$$+ \lambda e^{-\lambda n}E^P\left[\int_0^\infty e^{-\lambda t}f(X_n(n+t))dt\right].$$

At the same time,

$$E^P\left[\int_0^{1/n} e^{-\lambda t} f(X_n(\tfrac{m}{n}+t))dt\right]$$

$$= \int_0^{1/n} e^{-\lambda t} E^P[f(X_n(m/n)+\sigma(x(m/n))(\beta(t)-\beta(m/n)))]dt$$

$$= \int_0^{1/n} e^{-\lambda t} E^P\left[\int_{R^N} f(\sigma(x(m/n))y)g(t,y-X_n(m/n))dy\right]dt$$

$$\leq \int [R_\lambda f_m(\cdot,\omega)](X_n(m/n,\omega))P(d\omega)$$

where $f_m(y,\omega) \equiv f(\sigma(x(m/n,\omega))y)$. Note that, because of (5.1) and the fact that $\epsilon \leq 1/2$, there is a $K \in (0,\infty)$ such that $\|f_m(\cdot,\omega)\|_{L^p(\mathbb{R}^N)} \leq K\|f\|_{L^p(\mathbb{R}^N)}$ for all $m \geq 0$ and $\omega \in \Omega$. Hence, we have now proved that

$$E^P\left[\int_0^{1/n} e^{-\lambda t} f(X_n(\tfrac{m}{n}+t))dt\right] \leq KA\|f\|_{L^p(\mathbb{R}^N)};$$

and the same argument shows that

$$E^P\left[\int_0^{\infty} e^{-\lambda t} f(X_n(n+t))dt\right] \leq KA\|f\|_{L^p(\mathbb{R}^N)}.$$

Combining these, we conclude that $\int f d\mu_n \leq (n^2+1)KA\|f\|_{L^p(R^N)}$ for all $f \in \mathcal{S}(R^N)^+$.

Q.E.D.

With Lemma (5.10), we have now completed the proof of the following theorem.

(5.11)<u>Theorem</u>: Let $a: \mathbb{R}^N \longrightarrow S^+(\mathbb{R}^N)$ be a bounded continuous function satisfying $a(x) > 0$ for each $x \in \mathbb{R}^N$. Then, for every bounded measurable $b: [0,\infty)\times\mathbb{R}^N \longrightarrow \mathbb{R}^N$, the martingale problem for the associated L is well-posed.

(5.12)<u>Remark</u>: There are several directions in which the preceding result can be extended. In the first place, if a depends on $(t,y) \in [0,\infty)\times\mathbb{R}^N$ in such a way that for each $T > 0$ and $K \subset\subset \mathbb{R}^N$ the family $\{a(t,\cdot)\restriction K: t\in[0,T]\}$ is uniformly positive and equicontinuous, then the martingale problem for a and any bounded measurable b is well-posed. Moreover, when $N = 1$, this continues to be true without any continuity assumption on a, so long as a is uniformly positive on compact subsets of $[0,\infty)\times\mathbb{R}^N$. Finally, if $N = 2$ and a is independent of t, then the martingale problem for a and any bounded measurable b is well-posed so long as a is a bounded measurable function of x which is uniformly positive on compact subsets of $\mathbb{R}^N$. All these results can be proved by variations on the line of reasoning which we have presented here. For details, see 7.3.3 and 7.3.4 on pages 192 and 193 of [S.&V.].

Appendix:

In this appendix we will derive the estimate (5.6). There are various approaches to such estimates, and some of these approaches are suprisingly probabilistic. In order to give a hint about the role that probability theory can play, we will base our proof on Burkholder's inequality. However, there is a bit of preparation which we must make before we can bring Burkholder's inequality into play.

Given $1 \le j \le N$, define $\mathcal{R}_j$ on $\mathcal{S}(\mathbb{R}^N)$ to be the operator given by $\mathcal{F}\mathcal{R}_j f(\xi) = (i\xi_j/|\xi|)\mathcal{F}f(\xi)$. (Recall that we use $\mathcal{F}$ to denote the Fourier transform.) Then (5.6) comes down to showing that for each $p \in (1,\infty)$ there is a $C_p < \infty$ such that

$$\max_{1\le j\le N} \|\mathcal{R}_j f\|_{L^p(\mathbb{R}^N)} \le C_p \|f\|_{L^p(\mathbb{R}^N)}, \quad f \in \mathcal{S}(\mathbb{R}^N). \tag{A.1}$$

Indeed, suppose that (A.1) has been proved. Then we would have that $\|\partial_{y_j}\partial_{y_{j'}} f\|_{L^p(\mathbb{R}^N)} = \|\mathcal{R}_j\mathcal{R}_{j'}\Delta f\|_{L^p(\mathbb{R}^N)} \le C_p^2\|\Delta f\|_{L^p(\mathbb{R}^N)}$, since $\mathcal{F}\partial_{y_j}\partial_{y_{j'}} f(\xi) = -\xi_j\xi_{j'}\mathcal{F}f(\xi) = (\xi_j\xi_{j'}/|\xi|^2)\mathcal{F}\Delta f(\xi) = -\mathcal{F}\mathcal{R}_j\mathcal{R}_{j'}f(\xi)$.

(A.2) Lemma: There is a $c_N \in \mathbb{C}$ such that, for each $1 \le j \le N$, the mapping

$$\varphi \in \mathcal{S}(\mathbb{R}^N) \longmapsto \lim_{\epsilon\downarrow 0} \int_{|x|>\epsilon} c_N(x_j/|x|^{N+1})\varphi(x)dx$$

determines the tempered distribution T_j whose Fourier transform is $i(\xi_j/|\xi|)$ ($\equiv 0$ if $\xi = 0$). In particular, $\mathcal{R}_j f = f*T_j$, $f \in \mathcal{S}(\mathbb{R}^N)$ (again, "*" is used to denote convolution).

Proof: Note that for $\varphi \in \mathcal{S}(\mathbb{R}^N)$:

$$\int_{|x|>\epsilon} (x_j/|x|^{N+1})\varphi(x)dx =$$

$$\int_{\epsilon<|x|\le 1} (x_j/|x|^{N+1})(\varphi(x)-\varphi(0))dx + \int_{|x|>1} (x_j/|x|^{N+1})\varphi(x)dx$$

and

$$\lim_{\epsilon\downarrow 0}\int_{\epsilon<|x|\le 1} (x_j/|x|^{N+1})(\varphi(x)-\varphi(0))dx = \int_{|x|\le 1} (x_j/|x|^{N+1})(\varphi(x)-\varphi(0))dx.$$

Since

$$\varphi \in \mathcal{S}(\mathbb{R}^N) \longmapsto \int_{|x|\le 1} (x_j/|x|^{N+1})(\varphi(x)-\varphi(0))dx + \int_{|x|>1} (x_j/|x|^{N+1})\varphi(x)dx$$

is obviously a continuous linear functional on $\mathcal{S}(\mathbb{R}^N)$, we will have completed the proof once we show that:

$$\lim_{\epsilon\downarrow 0}\lim_{R\uparrow\infty}\int_{\epsilon<|x|\le R} (x_j/|x|^{N+1})\exp(i\xi\cdot x)dx = c\xi_j/|\xi|$$

for some $c \in \mathbb{C}\backslash\{0\}$. Clearly there is nothing to do when $\xi = 0$. When $\xi \neq 0$, standard calculations show that

$$\lim_{\epsilon\downarrow 0}\lim_{R\uparrow\infty}\int_{\epsilon<|x|\le R} (x_j/|x|^{N+1})\exp(i\xi\cdot x)dx$$

$$= \lim_{\epsilon\downarrow 0}\lim_{R\uparrow\infty} i\int_{S^{N-1}}\omega_j d\omega\int_{\epsilon<r\le R}\sin(r(\xi\cdot\omega))\frac{dr}{r} = \frac{i\pi}{2}\int_{S^{N-1}}\omega_j \operatorname{sgn}(\xi\cdot\omega)d\omega.$$

At the same time, if $(e_1',\ldots,e_N')$ is an orthonormal basis of $\mathbb{R}^N$ with $e_1' = \xi/|\xi|$, then

$$\int_{S^{N-1}}\omega_j\operatorname{sgn}(\xi\cdot\omega)d\omega = \sum_{\nu=1}^{N}(e_\nu'\cdot e_j)\int_{S^{N-1}}(e_\nu'\cdot\omega)\operatorname{sgn}(e_1'\cdot\omega)d\omega$$

$$= (e_j\cdot\frac{\xi}{|\xi|})\int_{S^{N-1}}|(e_1'\cdot\omega)|d\omega = k(\xi_j/|\xi|)$$

where $k \in \mathbb{C}\backslash\{0\}$ and is independent of $1 \le j \le N$. (In the preceding and in what follows, $(e_1,\ldots,e_N)$ is used to denote the standard ortho-normal basis in $\mathbb{R}^N$.)

Q.E.D.

Now define $r_j(x) = c_N(x_j/|x|^{N+1})$ on $\mathbb{R}^N\backslash\{0\}$; and, for $\epsilon > 0$, set $r_j^{(\epsilon)}(x) = \chi_{[\epsilon,\infty)}(|x|)r_j(x)$ and define $\mathcal{R}_j^{(\epsilon)}f = f*r_j^{(\epsilon)}$, $f \in \mathcal{S}(\mathbb{R}^N)$. In view of Lemma (A.2), what we must show is that for each $p \in (1,\infty)$ there is a $C_p < \infty$ such that

$$\sup_{\epsilon>0} \|\mathcal{R}_j^{(\epsilon)}f\|_{L^p(\mathbb{R}^N)} \le C_p\|f\|_{L^p(\mathbb{R}^N)}, \quad f \in \mathcal{S}(\mathbb{R}^N). \tag{A.3}$$

To this end, note that

$$\begin{aligned}\mathcal{R}_j^{(\epsilon)}f(x) &= c_N \int_{S^{N-1}} \omega_j d\omega \int_\epsilon^\infty f(x-r\omega)\frac{dr}{r} \\ &= (c_N/2)\int_{S^{N-1}} \omega_j d\omega \int_{|r|>\epsilon} f(x-r\omega)\frac{dr}{r}.\end{aligned}$$

Next, choose a measurable mapping $\omega \in S^{N-1} \longmapsto U_\omega \in O(N)$ so that $U_\omega e_1 = \omega$ for all ω; and, given $f \in \mathcal{S}(\mathbb{R}^N)$, set $f_\omega(y) = f(U_\omega y)$. Then:

$$\begin{aligned}\mathcal{R}_j^{(\epsilon)}f(x) &= (c_N/2)\int_{S^{N-1}} \omega_j d\omega \int_{|r|>\epsilon} f_\omega(x-re_1)\frac{dr}{r} \\ &= (c_N/2\pi)\int_{S^{N-1}} \omega_j \mathcal{H}^{(\epsilon)}f_\omega(x)d\omega\end{aligned}$$

where

$$\mathcal{H}^{(\epsilon)}g(x) \equiv (1/\pi)\int_{|r|>\epsilon} g(x-re_1)\frac{dr}{r}, \quad g \in \mathcal{S}(\mathbb{R}^N).$$

In particular, set $h^{(\epsilon)}(x) = \chi_{[\epsilon,\infty)}(|x|)\frac{1}{x}$, $x \in \mathbb{R}^1$, and suppose that we show that

$$\sup_{\epsilon>0} \|\psi*h^{(\epsilon)}\|_{L^p(\mathbb{R}^N)} \le K_p\|\psi\|_{L^p(\mathbb{R}^N)}, \quad \psi \in \mathcal{S}(\mathbb{R}^1), \tag{A.4}$$

for each $p \in (1,\infty)$ and some $K_p < \infty$. Then we would have that $\sup_{\epsilon>0} \|\mathcal{H}^{(\epsilon)}f_\omega\|_{L^p(\mathbb{R}^N)} \le K_p\|f_\omega\|_{L^p(\mathbb{R}^N)} = K_p\|f\|_{L^p(\mathbb{R}^N)}$ for all $\omega \in S^{N-1}$; and so, from the preceding, we could conclude that (A.3) holds with $C_p = |c_N|K_p/2\pi$. In other words, everything reduces to proving (A.4). (Note that the preceding reduction allows us to obtain (A.3) for arbitrary $N \in Z^+$ from the case when $N = 1$.)

For reasons which will become apparent in a moment, it is better to replace the kernel $h^{(\epsilon)}$ with $h_\epsilon(x) = \frac{1}{\pi}\frac{x}{x^2+\epsilon^2}$. Noting that

$$\pi\|h^{(\epsilon)}-h_\epsilon\|_{L^1(\mathbb{R}^1)} \le 2\int_0^\epsilon x/(x^2+\epsilon^2)dx + 2\int_\epsilon^\infty \epsilon^2/(x(x^2+\epsilon^2))dx$$
$$= 2\int_0^1 x/(x^2+1)dx + 2\int_1^\infty 1/(x(x^2+1))dx,$$

we see that (A.4) will follow as soon as we show that

$$\sup_{\epsilon>0} \|\psi * h_\epsilon\|_{L^p(\mathbb{R}^1)} \le K_p\|\psi\|_{L^p(\mathbb{R}^1)}, \quad \psi \in \mathcal{S}(\mathbb{R}^1),$$

for each $p \in (1,\infty)$ and some $K_p < \infty$. In addition, because:

$$\int\varphi(x)(\psi * h_\epsilon)(x)dx = -\int\psi(y)(\varphi * h_\epsilon)(y)dy, \quad \varphi,\psi \in \mathcal{S}(\mathbb{R}^1),$$

an easy duality argument allows us to restrict our attention to $p \in [2,\infty)$; and therefore we will do so.

Set $p_y(x) = \frac{1}{\pi}\frac{y}{x^2+y^2}$ for $(x,y) \in \mathbb{R}^2_+ \equiv \{(x,y) \in \mathbb{R}^2: y > 0\}$. Given $\psi \in C_0^\infty(\mathbb{R}^1;\mathbb{R}^1)$ (we have emphasized here that ψ is real-valued), define $u_\psi(x,y) = \psi * p_y(x)$ and $v_\psi(x,y) = \psi * h_y(x)$.

(A.6) <u>Lemma</u>: Referring to the preceding, u_ψ and v_ψ are conjugate harmonic functions on $\mathbb{R}^2_+$ (i.e. they satisfy the Cauchy-Riemann equations). Moreover, there is a $C = C_\psi < \infty$ such that $|u_\psi(x,y)|\vee|v_\psi(x,y)| \le C/(x^2+y^2)^{1/2}$ for all $(x,y) \in \mathbb{R}^2_+$. Finally, $\lim_{\delta\downarrow 0}\sup_x \sup_{|\xi-x|\vee y<\delta}|u_\psi(\xi,y)-\psi(\xi)| = 0$.

<u>Proof</u>: Set $F(z) = u_\psi(x,y) + iv_\psi(x,y)$ for $z = x + iy$ with

$(x,y) \in \mathbb{R}^2_+$; and note that $F(z) = \frac{i}{\pi} \int_{-\infty}^{\infty} \frac{\psi(\xi)}{z-\xi}\, d\xi$. Clearly, all the assertions about u_ψ and v_ψ, except the last one about u_ψ, follow immediately from the preceding representation of F. On the other hand, the asserted behavior of $u_\psi(\cdot,y)$ as $y\downarrow 0$ can be easily derived from elementary estimates. Q.E.D.

We are at last ready to see how Burkholder's inequality enters the proof of (A.5). Namely, for $(x,y) \in \mathbb{R}^2_+$, let $\mathscr{W}_{x,y}$ denote the two-dimensional Wiener measure starting from (x,y). Using $z(t,\omega) = (x(t,\omega),y(t,\omega))$ to denote the position $\omega(t) \in \mathbb{R}^2$ of the path $\omega \in \Omega$ at time $t \geq 0$, define $\tau_\epsilon(\omega) = \inf\{t\geq 0: y(t,\omega)\leq\epsilon\}$. We must first check that $\mathscr{W}_{x,y}(\tau_\epsilon<\infty) = 1$ for all $0 \leq \epsilon \leq y$; and, obviously, it is enough to do so in the case when $\epsilon = 0$. But, by Itô's formula,

$$(\exp[-\lambda(t\wedge\tau_0)-(2\lambda)^{1/2}y(t\wedge\tau_0)], \mathscr{M}_t, \mathscr{W}_{x,y})$$

is a martingale for each $\lambda \geq 0$; and so

$$\begin{aligned}\exp[-(2\lambda)^{1/2}y] &= E^{\mathscr{W}_{x,y}}\left[\exp[-\lambda(t\wedge\tau_0)-(2\lambda)^{1/2}y(t\wedge\tau_0)]\right] \\ &\leq E^{\mathscr{W}_{x,y}}\left[\exp[-\lambda(t\wedge\tau_0)]\right].\end{aligned}$$

Hence,

$$\begin{aligned}\mathscr{W}_{x,y}(\tau_0<\infty) &= \lim_{\lambda\downarrow 0} E^{\mathscr{W}_{x,y}}\left[\exp[-\lambda\tau_0]\right] \\ &= \lim_{\lambda\downarrow 0}\lim_{t\uparrow\infty} E^{\mathscr{W}_{x,y}}\left[\exp[-\lambda(t\wedge\tau_0)]\right] \geq \lim_{\lambda\downarrow 0} \exp[-(2\lambda)^{1/2}y] = 1.\end{aligned}$$

Next, let $\psi \in C_0^\infty(\mathbb{R}^1)$ be given and define

$$M_\epsilon(t,\omega) = u_\psi(z(t\wedge\tau_\epsilon(\omega),\omega)) - u_\psi(x,y)$$

and

$$N_\epsilon(t,\omega) = v_\psi(z(t\wedge\tau_\epsilon(\omega),\omega)) - v_\psi(x,y),$$

where u_ψ and v_ψ are defined relative to ψ as in the preceding. Then, by Itô's formula and Lemma(A.6), $(M_\epsilon(t),\mathcal{M}_t,\mathcal{W}_{x,y})$ and $(N_\epsilon(t),\mathcal{M}_t,\mathcal{W}_{x,y})$ are bounded martingales for each $(x,y) \in R^2_+$ and $0 < \epsilon \le y$. In addition, since (by the Cauchy-Riemann equations) $|\nabla u_\psi| = |\nabla v_\psi|$, we have

$$\langle N_\epsilon\rangle(t) = \int_0^{t\wedge\tau_\epsilon} |\nabla u_\psi(z(s))|^2 ds = \int_0^{t\wedge\tau_\epsilon} |\nabla v_\psi(z(s))|^2 ds = \langle M_\epsilon\rangle(t)$$

(a.s.,$\mathcal{W}_{x,y}$). Hence, by Burkholder's inequality (cf. (3.18)), we see that for each $p \in [2,\infty)$ there is a $C_p < \infty$ such that

$$\|N_\epsilon(\tau_\epsilon)\|_{L^p(\mathcal{W}_{x,y})} \le \|N^*_\epsilon(\tau_\epsilon)\|_{L^p(\mathcal{W}_{x,y})} \le C_p\|M^*_\epsilon(\tau_\epsilon)\|_{L^p(\mathcal{W}_{x,y})}$$
$$\le C_p\|M^*_0(\tau_0)\|_{L^p(\mathcal{W}_{x,y})} \le (p/(p-1))K_p\|M_0(\tau_0)\|_{L^p(\mathcal{W}_{x,y})},$$

where we have used Doob's inequality in order to get the last relation. Since $y(\tau_\epsilon) = \epsilon$ (a.s.,$\mathcal{W}_{x,y}$), we conclude from the preceding that

$$E^{\mathcal{W}_{x,y}}\Big[|v_\psi(x(\tau_\epsilon),\epsilon)-v_\psi(x,y)|^p\Big]^{1/p}$$
$$\le K_p E^{\mathcal{W}_{x,y}}\Big[|\psi(x(\tau_0))-u_\psi(x,y)|^p\Big]^{1/p};$$

and so

$$E^{\mathcal{W}_{x,y}}\Big[|v_\psi(x(\tau_\epsilon),\epsilon)|^p\Big]^{1/p}$$
$$\le K_p E^{\mathcal{W}_{x,y}}\Big[|\psi(x(\tau_0))|^p\Big]^{1/p} + |u_\psi(x,y)| + K_p|v_\psi(x,y)|$$

for all $(x,y) \in \mathbb{R}^2_+$ and $0 < \epsilon \le y$. Noting that the distribution of $x(\tau_\epsilon)$ under $\mathcal{W}_{x,y}$ is the same as that of $x+x(\tau_\epsilon)$ under $\mathcal{W}_{0,y}$, raising the preceding to the power p, and integrating with respect to $x \in R^1$, we obtain:

$$\|v_\psi(\cdot,\epsilon)\|^p_{L^p(\mathbb{R}^1)} \leq 2^{p-1}K_p^p\|\psi\|^p_{L^p(\mathbb{R}^1)} + 2^{p-1}\left[\|u_\psi(\cdot,y)\|^p_{L^p(\mathbb{R}^1)} + K_p^p\|v_\psi(\cdot,y)\|^p_{L^p(R^1)}\right]$$

for all $0 < \epsilon \leq y$. But, by the estimate on u_ψ and v_ψ in Lemma(A.6), it is clear that $\|u_\psi(\cdot,y)\|^p_{L^p(R^1)} \vee \|v_\psi(\cdot,y)\|^p_{L^p(R^1)} \longrightarrow 0$ as $y\uparrow\infty$. In other words, we have now proved (A.5), with $2K_p$ replacing K_p, so long as $\psi \in C_0^\infty(\mathbb{R}^1)$. Obviously, the same result for all $\psi \in \mathcal{S}(\mathbb{R}^1)$ follows immediately from this; and so we are done.

INDEX